THE SCIENCE OF ANIMAL WELFARE

The Science of Animal Welfare

Understanding What Animals Want

Marian Stamp Dawkins

Department of Zoology
University of Oxford, UK

OXFORD
UNIVERSITY PRESS

OXFORD
UNIVERSITY PRESS

Great Clarendon Street, Oxford, OX2 6DP,
United Kingdom

Oxford University Press is a department of the University of Oxford.
It furthers the University's objective of excellence in research, scholarship,
and education by publishing worldwide. Oxford is a registered trade mark of
Oxford University Press in the UK and in certain other countries

First Edition published in 2021
Impression: 2

Published in the United States of America by Oxford University Press
198 Madison Avenue, New York, NY 10016, United States of America

British Library Cataloguing in Publication Data
Data available

Library of Congress Control Number: 2020949499

ISBN: 978–0–19–884898–1 (hbk.)
978–0–19–884899–8 (pbk.)

DOI: 10.1093/oso/9780198848981.001.0001

Printed and bound by
CPI Group (UK) Ltd, Croydon, CR0 4YY

The welfare of animals must depend on an understanding of animals and one does not come by that understanding intuitively: it must be learned.

P. B. Medawar (1972) *The Hope of Progress*

I would like to thank Jonathan Kingdon for allowing me to use part of one of his paintings for the cover of this book. The full painting shows the red eye and striking plumage of a male vulturine guinea fowl as a female might see him, with the margins suggesting how adult plumage derives from juvenile camouflage.

Contents

Preface

I wrote this book for two reasons. One was to clarify what is meant by animal welfare in a way that would be accessible to anyone, whatever their views on animals and whether they are scientists or non-scientists. The other was to put an animal's point of view at the centre of how we assess their welfare.

The book is not intended either as a textbook or as a campaigning book. It is more a guide for anyone who is interested in animals and how their welfare can best be assessed scientifically. There are full references to the scientific literature so I hope that it will be useful to scientists and to students, particularly in biology and veterinary medicine, but I also hope that it is self-contained enough to be clear to everyone else, whatever their background or previous knowledge. My aim is to show how science can be used to discover what is best for animal welfare, but to do so in a way that leaves it up to each individual reader to decide for themselves how the facts we have discovered should be used to change, or not change, the way animals are treated.

I would like to thank numerous colleagues for discussions that have helped in the writing of this book, including Christine Nicol, Sabine Gebhardt-Henrich and Edmund Rolls. Conversations with David Wood-Gush are still vividly remembered.

One small point. I have used the term 'animal' throughout the book to mean 'non-human animal'. Of course we humans are animals too, but to keep saying 'non-human' gets tedious after a while and makes sentences clumsy. So please take the term 'animal' as a convenient shorthand, not as a statement about our wider relationship to the rest of the animal kingdom.

M.S.D.

Oxford
October 2020

1

Animal Welfare: The Science and Its Words

If you are human, you cannot help being touched in some way by the other animals that share this planet with us. Even if you are unaware of them, they affect your health, the food you eat, the medicines you take. And if you interact with them in any way, you will know how profoundly animals can also affect our lives as companions, pests and sources of great wonder and beauty.

The welfare of different species—and no-one knows exactly how many of them there are—is of increasing importance to many people. In response to rising public concern, the past 40 years has seen an unprecedented growth in laws and regulations to protect animals and, more positively, to give them better lives.

However, if these laws and regulations are really to achieve their aim of improving the lives of animals, they need to be evidence-based—that is, based on what can actually be shown to improve animal welfare rather than on what well-meaning people imagine might improve it. Different animal species are different—different from us and different from each other. We all share the same needs for basics such as nutrition and shelter, but there are vast differences in how these needs are met because of our differences in lifestyle, environment and genetics. These differences need to be researched and understood if good welfare and quality of life are to be achieved for all species.

That is where the science of animal welfare comes in. It is a science dedicated to providing an evidence-based approach to animal welfare. But it is a science with some rather peculiar problems of its own, arising partly out of its subject matter (what is animal welfare?) and partly out of its frequent use of emotionally laden words such as 'suffering', 'fear' and 'stress' that carry with them built-in views about what is good or bad. Unlike, say, geology, which can be defined as the study of the Earth and its rocks without expressing any opinion about what ought to be done about either the Earth or its rocks, animal welfare science is deeply enmeshed in the power of its own words. The word 'welfare' itself expresses the view that achieving good welfare is a desirable aim, and calling an animal 'fearful', 'frustrated' or even just 'restricted' presupposes the conclusion that what is happening to the animal must be bad for it and should be stopped. Just think of the difference that is implied by describing an animal as 'calm' and describing that same animal as 'inactive' or 'inert' and you can see the power of words to

The Science of Animal Welfare: Understanding What Animals Want. Marian Stamp Dawkins, Oxford University Press (2021).
 DOI: 10.1093/oso/9780198848981.003.0001

completely alter our view of a situation and give a very different steer as to what ought to be done about it.

So the words we use to describe animals and their welfare are not neutral. They come loaded with emotional overtones of what is good or bad. They are full of their own implications of what ought to be done and subtly encourage the kind of thinking (all too easy to slip into anyway) that bypasses careful consideration of what is actually happening and what scientific evidence we need to gather.

This book is an attempt to cut through the words and unclear definitions that often confuse thinking about animal welfare and to lay out a truly scientific, animal-centred way of defining good welfare. We will see that it is possible to define animal welfare in an objective way that can be easily applied and understood (Chapters 2 and 3) and to use this definition to evaluate the different kinds of information we now have to make evidence-based decisions about how to improve it (Chapters 4, 5, 6 and 7). The thread running through the book is the importance of taking into account what animals themselves want, so that welfare is viewed not just from a human perspective but from the animal's point of view too.

Making what animals want the very heart of what 'welfare' is allows us to bring together many years of research by many people and to show how apparently different ideas fit together. As we will see, it also allows us to resolve many of the controversies that have arisen about what are valid or invalid measures of welfare. In that sense, the book sets out to be a straightforward look at animal welfare science today.

There is another sense, however, in which it which it departs quite radically from some current definitions, particularly those that define animal welfare primarily in terms of conscious experience. For the reasons explained in Chapter 2, the definition of 'animal welfare' we will be using rests entirely on what we can actually observe and measure in practice—that is, on how animals behave and their physiology. It does not mention, at least in the first instance, sentience or what animals are consciously feeling. Defining welfare without consciousness does not, of course, deny animal sentience or imply that animals lack conscious experiences. It simply means—temporarily—leaving the issue to one side on the grounds that consciousness is itself so difficult to study that including it in the definition of welfare only leads to confusion and controversy. In fact, as we will see in Chapter 2, one of the main reasons why there is currently no generally agreed definition of welfare is because there is no generally agreed definition of consciousness. A definition of welfare that does not mention conscious feelings can therefore be thought of as an intellectual safety device, a way of avoiding being distracted by terms that are difficult to define or emotionally loaded. Using this device, we can arrive safely at testable, objective hypotheses about what animal welfare is and how it could be improved. Then, in Chapter 8, we throw off the safety harness and look at what we now know about animal sentience and its role in animal welfare.

First, however, we need to look at the biggest of all word problems that animal welfare science faces—a definition of what welfare is.

2

What Is Animal Welfare?

Fifty years ago there was no recognized science of animal welfare. There was just a collection of vets, ethologists, geneticists and other people scattered around the world who were linked by the belief that animal welfare was important and deserved to be taken seriously as a science in its own right. With such diverse starting points, a single definition of 'welfare' was unlikely to emerge easily.

What is more surprising is that now, with animal welfare science an established discipline, with its own journals and textbooks and international societies, there is still no agreed definition of 'welfare' or a consensus on how to improve it (Green and Mellor 2011; Ede et al. 2019; Weary and Robbins 2019; Polgár et al. 2019). Some people, for example, argue is that the only way to guarantee the welfare of an animal is to make its environment as 'natural' as possible, whereas others will claim that a natural life does not guarantee good welfare and that animals' needs can be better met in a controlled, if artificial, environment. Each side here is using a different definition of 'welfare', different methods for assessing it and coming up with a completely different answer as a result. About the only thing that commands a measure of universal agreement is that welfare is very complex and that assessing it requires evidence from many different sources (Mason and Mendl 1993; Fraser 2008; Mellor 2016a). But from a practical point of view, this is clearly not good enough. For something as important to many people as animal welfare, and certainly for trying to make improvements to the lives of animals, we need to know what welfare *is*, not just that it is difficult to define.

We therefore start our exploration of animal welfare (or 'well-being' as it is sometimes called) by trying to say exactly what it is we are talking about. Furthermore, this needs to be done in terms that everyone—farmers, vets, politicians, philosophers, scientists and the general public—can all understand and buy into. Animal welfare may now be a scientific discipline but it is one that touches the rest of the world very directly. People everywhere therefore want access to the important advances that are being made in understanding the worlds of animals.

In this chapter, we will see that there are two main reasons why people disagree about what the term 'animal welfare' should mean. One is the multiplicity of different ways that are now used to measure 'welfare', including physiology, health, hormone levels, behaviour, immunology and choice tests, which may all give conflicting answers. This leaves people unable to agree on which ones to rely on and which ones deserve top priority in the definition of 'good welfare'. The other is the widespread desire to put subjective

The Science of Animal Welfare: Understanding What Animals Want. Marian Stamp Dawkins, Oxford University Press (2021).
 DOI: 10.1093/oso/9780198848981.003.0002

experiences—that is, the way animals consciously experience pain, pleasure and suffering—at the core of the definition. Consciousness, however, is difficult enough to study in humans and even more so in other species, which means that making conscious feelings an essential part of the definition of welfare has inevitably lead to controversy. We can refer to these two issues as the *complexity* problem and the *consciousness* problem, respectively. Neither need prove fatal to a universally agreed view of what good welfare is, but before we can arrive at an agreed definition of animal welfare, we need to deal with each of them in turn.

The complexity problem

Animal welfare does not lack 'measures' of welfare. Among the many ways of measuring welfare that have been proposed we find: longevity, surface injuries, immune function, increase in activity, decrease in activity, H-index (a measure of behavioural diversity), sleep, play, stereotypies, exploration, response to novel objects, approach distance to humans, choice, grooming of self, grooming of others, telomere length, skin temperature, eye temperature, hormone levels, pupil size, cognitive bias, time to build a nest and running speed. There are many more. The problem is not too few measures of welfare but how to make sense of the many that are available and which ones can be most reliably used to define good welfare.

Early on in the development of animal welfare science, the first attempts to measure animal welfare acknowledged that this was indeed a very complex problem and that the best approach was to make as many different measures as possible and hope that a composite picture would somehow emerge (Dawkins 1980; Broom and Johnson 1993). This was a bit like being unable to get into a room but looking in through many different windows, as Jane Goodall (1990) put it. The different windows were things like measuring stress hormones, weighing adrenal glands, looking for unusual behaviour, assessing the animal's health and so on. It was soon realized, however, that these different measures did not always show the same thing (Mason and Mendl 1991) and that what seemed to be going on in the room depended on which window you looked through. For example, animals might show a rise in corticosteroid hormone (sometimes called 'stress' hormone) not just when they were in an obviously stressful situation such as being chased by a predator but also when they were anticipating food, having sex or given access to a preferred environment (Rushen 1991).

A widely adopted solution of how to deal with such 'contradictory' measures is to make as many different measurements as possible and then take a balanced consensus view of all of them. For example, Welfare Quality® (2009, 2018), a European-wide project with the ambitious aim of defining welfare for different farmed species, involves a series of detailed protocols for assessing four key welfare area of good housing, good feeding, good health and appropriate behaviour. For each area there are 50 or more measurements, which are then combined into a final score of 'excellent', 'enhanced', 'acceptable' or 'not classified' (Buijs et al. 2017). Welfare is thus defined as the weighted sum of its many component parts. Of course, the final answer depends critically on how much weight is given to each component, so the issue of how to combine or reconcile the different measures does not go away. The room still looks different depending on which window you look through most often.

So the first requirement for a universal definition of animal welfare is that it must somehow be able to accommodate the wide variety of different ways that people now have of measuring welfare. It must also justify the priority given to these different measures in arriving at the final combined answer of what good welfare actually is. But before we can do that, we have to deal with an even greater problem that has also got in the way of a universally agreed definition of animal welfare—the issue of what animals consciously *feel*.

The consciousness problem

The belief that non-human animals have the capacity to consciously feel pain and pleasure (often referred to as 'sentience') is what for many people distinguishes animal welfare from, say, the care of plants or the curation of valuable works of art (Singer 1976; Midgley 1983; Rollin 1989). The central importance of sentience in shaping attitudes to animals is often traced back to Jeremy Bentham's (1789) famous statement about dogs and horses: 'The question is not, Can they *reason*? nor Can they *talk*? but, Can they *suffer*?' More recently, Singer (1976) and other philosophers have argued that sentience—particularly the capacity to suffer—should be the deciding factor in our concern for animals, and many biologists have also argued for defining welfare in terms of what animals consciously feel (Dawkins 1990; Duncan 1993; Webster 1994; Fraser 2008; Broom 2014; Mellor 2019). Across the world sentience is now used as the basis of animal welfare legislation, a notable example being the European Union's (2009) Lisbon Treaty, which explicitly states that animals are sentient beings. Many people take the view that it is so obvious that non-human species consciously experience feelings of pleasure, pain and suffering that the same mixture of intuition, guesswork and a willingness to give the benefit of the doubt that we use to conclude that other people are conscious can be used, suitably adjusted for biological differences, to conclude the same for other species as well (e.g. Panksepp 1998, 2011; Balcombe 2006; Bekoff 2007; Urquiza-Haas and Kotrschal 2015).

But this growing assumption of sentience in other animals only makes it more difficult to agree on a definition of animal welfare that suits everyone. Consciousness is the most elusive and difficult to study of all biological phenomena (Koch 2004; Blackmore and Troscianko 2018). Even with our own consciousness, we still do not understand how the lump of nervous tissue that makes up our brain gives rise to private subjective experiences—such as a pain that hurts us, a feeling of cold that is unpleasant or a sensation of seeing a red light that feels like anything at all. And because we do not understand how the human brain makes us conscious, we do not know what to look for in other species to decide if they, too, have conscious experiences like us. Perhaps they do, but how would we know? And what if the 'feelings' of a fish were so different from those of a dog that we would find it difficult to bring them into the same definition?

The past 20 years have seen an explosion of interest in trying to solve what has come to be known as the hard problem of consciousness (Chalmers 1995) and there are now a number of widely held theories about how it might be possible to recognize when a human brain switches from unconscious to conscious processing, which we will discuss more fully in Chapter 8. We have brain imaging techniques that effectively show us what is going on inside a living, thinking brain and a far greater knowledge than ever before about what different parts of the brain are doing as we go about different tasks, recall memories and slip in and out of consciousness. All this new information should have made it easier to determine which other species have conscious experiences more or less like ours. It would seem obvious that the better we understand what brain mechanisms make us conscious, the better equipped we would be to judge whether other species have similar mechanisms, but in fact just the opposite seems to have happened.

Depending on which theory of consciousness you choose to believe, it is now possible find an extraordinary range of conclusions being drawn about which animals are conscious. At one extreme, we find claims that almost all of them are, to, at the other, that none of them are. Proposals for membership of the 'consciousness club' include that it is for humans only (Macphail 1987), for language users only (Rosenthal 1993, 2005), for humans and apes only (Bermond 2001), for all mammals (Boly et al. 2013), for mammals and birds (Seth et al. 2005), for mammals, birds and reptiles but not fish or amphibia (Cabanac et al. 2009), for all vertebrates including fish (Denton et al. 2009; Mashour and Alkire 2013; Braithwaite 2010; Sneddon 2019), for all vertebrates and a few invertebrates such as octopuses (Tye 2017), for many invertebrates especially insects and crustacea (Barron and Klein 2016; Bronfman and Ginsberg 2016) or that it should be for all living things, including plants (Margulis 2001). It has even been claimed that consciousness is everywhere, even in inanimate objects (Chalmers 2016; Kastrup 2018).

This lack of agreement among scientists about animal consciousness or even which animals are capable of having conscious feelings at all leaves the study of animal welfare at risk of looking vague, unscientific and unable to agree on its own core concept. Seeing what appears to be the inability of animal welfare science to understand animal consciousness, people outside the scientific community feel entitled to argue that their views are as good as anyone else's, including, perhaps particularly, those of scientists. The problem of how to understand what animals feel arouses strong and divergent opinions well outside animal welfare science itself. As a consequence, 'animal welfare' has come to means very different things to different people, united by the belief that it is about what animals feel but divided by how and even whether this can be studied scientifically.

Since finding a universally agreed definition of animal welfare has been made so difficult by the widespread desire to define it largely, if not exclusively, in terms of what animals feel, some animal welfare scientists have started to think that the best way to make progress is to use a definition that does not depend on conscious feelings at all (Arlinghaus et al. 2009; Würbel 2009; Dawkins 2012). This view does not deny consciousness to animals. On the contrary, it allows for the strong possibility that many of them do have vivid conscious experiences, perhaps very like our own, a possibility that we will discuss further in Chapter 8. All it says is that, for now, we need a definition of animal welfare that everyone, whatever their views on animal consciousness, can agree on. A possible way forward, therefore, is to define animal welfare without any mention of consciousness, sentience or subjective experiences whatsoever. In that way, everyone is free to have their own views about consciousness in other species, but consciousness does not form part of the definition of animal welfare itself. This has the important consequence that we do not have to have solved the hard problem of consciousness—the hardest problem in the whole of biology—before we can have a scientific study of animal welfare. Animal welfare without consciousness allows for the possibility of consciousness in other species but avoids the confusion and controversies that are created by trying to put conscious feelings into the definition itself (Dawkins 2015, 2017).

The two main reasons, then, why it has been so difficult to arrive at a definition of 'good welfare' that everyone can agree on are the complexity problem (so many different measures of welfare) and the consciousness problem (subjective feelings). We will now see that there is a relatively simple definition of animal welfare that can provide solutions to both of these problems and at the same time gives a solid framework for a scientifically based science of animal welfare. This definition has the ability to make sense of the many different 'measures' of welfare that are now in use and also avoids (while not denying) the possibility that animals have conscious experiences. It has the further advantage that it is very simple and so can be easily understood by scientists and non-scientists alike.

A basic definition of animal welfare

'Welfare' means literally 'going well' (it has the same two components as 'farewell'). In nature, success is measured in terms of survival and producing offspring, so an animal that is 'going well' is one that is not just alive now but is on course for still being alive in the future, at least for long enough to reproduce and pass its successful traits on to the next generation. The essence of good welfare is therefore being currently healthy and also having good prospects for future health. In that simple sentence lies the key to defining animal welfare.

Let us start with current health, which is universally accepted as the foundation of good animal welfare. For example, the Five Freedoms (Brambell 1965; Webster 1994), a widely used system for assessing welfare around the world, lists freedom from disease and injury as a key indicator of welfare. This is emphasized just as strongly in more recent versions such as the Ten General Principles (OIE 2012; Fraser et al. 2013), the Five Provisions or Domains (Mellor 2016a, 2016b) and the Four Principles put forward by the European Welfare Quality assessment (Welfare Quality® 2009). At least half of the criteria put forward by these and other welfare schemes are specifically aimed at maintaining animal health—such as making sure that animals have adequate food and water, and are kept in safe comfortable environments in which they are not injured. So there is no controversy over the importance of the current state of an animal's health to its welfare. We can, in line with all current thinking, list good health as the first part of the definition of animal welfare.

Indeed, many of the most outstanding welfare issues are regarded as serious precisely because they involve obvious physical injury and ill-health (Arlinghaus et al. 2009;

Würbel 2009; Dawkins 2012). For example, feather-pecking in laying hens (Gunnarsson et al. 1999) and tail-biting in pigs (Taylor et al. 2012) can lead to serious injury and even death. Death, injury and disease are clear health outcomes that can be measured in objective, scientific ways and so using them brings 'welfare' easily into the realm of scientific measurement and hypothesis testing.

However, most people would also argue that there is more to good welfare than just physical health and so, while current health status is important, it cannot, on its own, fully define welfare. Physical health tells us how well an animal's body is functioning now and perhaps about its likely health prospects for the future. It does not tell us how the animal itself is responding to the world around it, whether, for example, it is searching for something it cannot find (deprivation) or attempting to escape from something it cannot avoid. In other words, physical health alone does not give us the animal's own point of view (Dawkins 1990). It does not tell us what the animal itself wants.

'What the animal wants' is the second key component of a definition of animal welfare. Although it may sound a rather odd way of putting it, on closer examination 'what animals want' actually puts into understandable words what most people mean when they talk about good welfare. For example, if someone expresses concern about an animal kept in a zoo on the grounds that it is not free to carry out its natural behaviour, what they really mean is that the animal is unable to do many of the behaviours it wants to do and would do if it could. Or, if they describe a bird fluttering against the bars of its cage as 'suffering', what they mean is that here is a bird that wants to escape. Describing these situations in terms of what animals want (or do not want) avoids the pitfalls of using words like 'deprived', 'suffering' and so on that describe situations as we humans might see them. It asks the animals how they see things and points us clearly to how we can find out, as we will see in Chapter 4.

Using the very down-to-earth phrase 'what animals want' also helps us deal with the two problems with defining animal welfare that we discussed earlier in this chapter. It helps with the complexity problem because finding out what the animals themselves want allows us to validate a whole range of measures such as hormone levels, skin temperature or activity levels that, on their own, are difficult to interpret in welfare terms (Boissy et al. 2007; Dawkins 2008; Mendl et al. 2010). Finding out what the animals themselves want allows us to categorize these in terms of whether the animal regards a given situation as positive (to be approached and repeated) or negative (to be avoided). This positive–negative categorization is sometimes called 'valence' (Mendl et al. 2010) and, as we will see in the following chapters, is key to the correct interpretation of the different measures of welfare that have been proposed. Many questions, such as whether animals should be able to do all their natural behaviour or what level of stress hormone indicates poor welfare, immediately become much more tractable, once they are subjected to the test of what the animal itself wants. Positive valence is key to good welfare.

'What animals want' also provides a way of dealing with the consciousness problem that makes it so difficult for people with different views on animal sentience to agree on a definition of welfare. By being animal-centred but at the same time not making any assumptions, one way or the other, about conscious experiences in animals, it provides

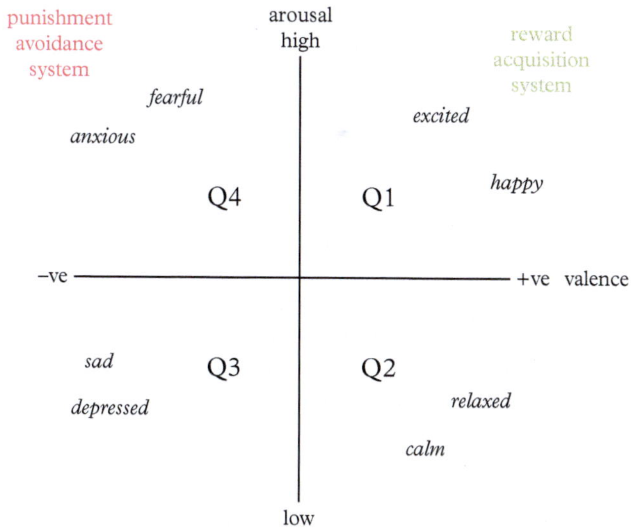

Figure 2.1 *How to put a variety of emotions into one diagram. The 'core affect' representation describes each emotion in two dimensions - valence (whether it is positive or negative) and intensity (how strong it is). The right half of the diagram shows positive emotions, associated with what is pleasurable or rewarding, while the left half shows negative emotions, associated with what is aversive or punishing. The core affect was originally developed to describe human emotions but is increasingly used in animal welfare science to describe emotions in animals. (Redrawn from Mendl et al. 2010).*

a unifying definition that most people can subscribe to. The value of this approach can best be illustrated by showing how 'what animals want' relates to one of the most important recent developments in animal welfare science, namely, the use of the 'affective state' framework (Mendl et al. 2010; Anderson and Adolphs 2014).

'Affective state' is a concept originally developed for describing human emotions (Russell and Barrett 1999; Russell 2003) and involves classifying all emotions along two dimensions: valence and arousal (Figure 2.1). Valence, as we have seen, refers to whether an emotion is positive or negative, so happiness and pleasure would have positive valence, while fear, anxiety and boredom would have negative valence (Figure 2.1). Arousal, on the other hand, indicates how strongly that emotion is felt—its intensity—and so can be a feature of either positive or negative emotions.

As applied to humans, affective state explicitly includes subjective feelings as part of the description of an emotion, so a positive emotion would be one that is consciously experienced as pleasurable, while a negative emotion such as fear or anger would be consciously experienced as aversive or unpleasant. However, in the human literature, it is also fully recognized that all emotions—positive and negative—have three separate components—the behavioural expression of that emotion, the physiological changes that occur, and the subjective feelings that may or may not accompany that behaviour and physiology (Keltner et al. 2013; Oatley and Jenkins 1996). Usually, of course, all three components occur together. When we get angry, we start shouting (behaviour), become red in the face (physiology) and feel furious (conscious experience). When we become fearful, we prepare to escape (behaviour), our hearts start racing (physiology) and we experience fear (subjective experience). But these components do not always go together and the conscious experience of an emotion can be, and often is, dissociated from the behavioural and physiological changes that normally accompany it. For example, fear-reducing drugs that alter the way people behave do not necessarily make them subjectively feel any less fearful (Le Doux 2014). People on such medication can appear calmer on the outside and even show a reduction in heart rate and other physiological indicators of fear but they still say they feel as anxious and fearful as ever (Le Doux and Hofmann 2018). This appears to be because, in humans, there are two separate brain circuits involved in fear: one involving the amygdala that controls the behavioural and physiological response to threats, and a completely separate cortical circuit that underlies the conscious experience of fear. Conscious feelings of fear do not require the amygdala (Anderson and Phelps 2002), and medication that targets the amygdala does not necessarily relieve subjective feelings of fear (LeDoux and Pine 2016; LeDoux and Hofmann 2018).

But difficult though it is to make the distinction between the behavioural, physiological and subjective components of emotion in humans, when it comes to applying the same ideas to non-human animals, the distinction is often lost altogether. Very confusingly, different researchers differ on whether or not they are talking about conscious feelings. For example, Mendl et al. (2010) carefully say that the concepts of affect, emotion and mood do not imply anything, one way or the other, about conscious feelings when transferred to animals. They write: 'Of course, even if we can use measurable components of emotional responses to locate an animal's position in core affect space, we cannot be certain that they experience the *conscious* components too.' They then go on to describe animal emotions as states that 'may or may not be experienced consciously' (p. 2896).

On the other hand, Ede et al. (2019) begin a recent review of cattle welfare by saying 'we use the words "affect", "emotion" and "feeling" synonymously', thus implying that for them 'affective state' does imply conscious feelings. Many papers on animal welfare use words such as 'emotion', 'optimism', 'pessimism', 'fear' and 'anxiety' in ways that also blur the distinction between observable behaviour and private subjective feeling, leaving the reader unclear whether conscious feelings are or are not implied. Even with disclaimers, the words that are used still carry with them the message of conscious experience because in everyday language, that is the message they are used to convey (LeDoux 2014).

The ambiguity about conscious feelings that is, often quite inadvertently, introduced into discussions of animal welfare by the use of the affective state language can be easily overcome by substituting 'having what it wants' for an animal with positive affect and 'not having what it wants' or 'having what it does not want' for negative affect. 'Valence' is directly equivalent to 'what the animal wants', but 'valence' too can be used in different ways, so tends to increase rather than to reduce ambiguity. Whereas Mendl et al. (2010) say that valence does not necessarily imply any subjective experiences, Webb et al. (2019)

explicitly equate valence with 'subjective experience of pleasantness or unpleasantness' (p. 62). Furthermore, if you say 'valence' to someone who is not a scientist, they will probably not know what you are talking about. Say 'what the animal wants' and they will immediately see the point. 'Valence' sounds good but what it actually means is what the animal itself wants.

There is one further point that needs to be clarified. The term 'wanting' is already used in a quite specific way by people who study the way animals learn. When an animal successfully learns a task, such as pressing a lever to get a reward of a piece of food, a whole chain of events happens (Berridge and Robinson 1998). First, the animal has to 'want' something it does not have—that is, it must be motivated to obtain the food. Second, the animal has to 'like' what it gets—that is, it must find the food pleasurable or it would not want to repeat the experience. Third, it has to learn how to get what it wants by associating its own behaviour with the subsequent reward of obtaining the food. Usually, of course, all this happens smoothly as a unified process of learning a new task. Animals 'want' what they 'like' and 'like' what they 'want'. But not always. We all know that it is possible to want something and then not like it when you get it and, as confirmation of this intuition, it is now known that 'wanting' and 'liking' are in fact associated with different brain mechanisms (Anselme and Robinson 2016). However, for the purposes of this book, 'what animals want' will be taken to include both. In other words, an animal that has 'what it wants' is one that is in a state where it 'likes' rather than 'dislikes' its current situation (Boissy et al. 2007; Webb et al. 2019) and is not highly motivated to obtain something it wants but cannot have (or get away from something it wants to avoid) (Dawkins 1990; Gygax 2017). As we will see in subsequent chapters, we can then establish what animals want by discovering what they find rewarding (i.e. what they like enough to learn to obtain) or punishing (i.e. what they dislike enough to learn to avoid). Thus, although 'liking' and 'wanting' are distinct aspects of reward, in the context of animal welfare, it is important to establish both that animals have what they want and that they like what they have.

In the rest of this book, we will use the definition of animal welfare as being a state where an animal is both healthy and has what it wants, to include liking what it has (Dawkins 2008; Gygax 2017). By applying this definition to different areas of animal welfare science, we will see that it has the ability to provide a meaning to 'good welfare' that can be easily understood by everyone, provides a simple but comprehensive guide to how to improve animal welfare in practice and can accommodate a wide range of views on consciousness in non-human animals. It thus aims to be the universally agreed definition of welfare that animal welfare science has so far been missing.

How 'health and what animals want' relates to other definitions

'Health and what animals want' is not a new definition of welfare. It is a condensed version of many current definitions that are already in use. For example, the first three of the Five Freedoms (FAWC 1979, 2009) refer to keeping animals healthy—freedom from hunger and thirst, freedom from discomfort, and freedom from injury and disease—while the remaining two refer to giving them what they want. Freedom 4 is freedom to perform most normal behaviour and Freedom 5 is freedom from stress. The implication behind these last two freedoms is that animals want to perform their natural behaviour and do not want to be stressed, which can both be tested by finding out what the animals themselves really want. Welfare Quality® (2009) lists 'appropriate behaviour' as one of its Welfare Principles, leaving a question over what is meant by 'appropriate', which can be best answered by asking the animals themselves what it is they want to do. All that the definition of 'health and what animals want' does is to clarify what many people have already said about welfare and to point clearly in the direction of what needs to be done to improve it.

'What animals want' is also in line with the recent trend to move away from defining welfare negatively as absence of suffering to defining it more positively (Boissy et al. 2007) so that animals have a life worth living (LWL) or, even better, a positively 'good life' (FAWC 2009; Wathes 2010; Green and Mellor 2011; Webb et al. 2019). What constitutes a good life from an animal's point of view can be made real by practical research on what they need to keep them healthy but also on what they demonstrate to us that they want. It can even be used by those who have suggested that animal welfare is not just a scientific concept but should also include keeping animals in ways that the public believes they should be kept (McInerney 1991; Yeates 2017). Practical information on what keeps animals healthy and what they want is the best way to ensure that animal interests are not overruled by well-meaning but erroneous public views on how animal should be kept.

Conclusions

In this chapter, we have seen that many of the problems of defining welfare come partly from the fact that there are now so many different 'measures' of welfare and partly from persistent difficulties of trying to incorporate conscious feelings into the definition. We have also seen, however, that defining welfare in terms of two basic elements of welfare—health and what animals want—is a way of addressing both of these problems by focusing on what is measurable and what needs to be tested. Defining animal welfare without explicitly including subjective feelings does not preclude the possibility that one day, when we understand consciousness better than we do now, we can then include conscious feelings in an ultimate definition of welfare, and it certainly does not rule out the idea that many non-human animals are able to consciously experience pain, pleasure, frustration and a host of other positive and negative emotional states. It just points out that trying to include in a definition of welfare something as elusive and little-understood as conscious feelings has led to confusion and an inability to arrive at a definition of welfare on which everyone can agree. This is not good for animal welfare science, and it is not good for animals or the prospects of improving their welfare. It is important to start with what we can measure and what we can agree on.

'Health and what animals want' also has the advantage that it includes at least part of what everyone, both inside and outside science, already seems to mean by welfare even though they may not put it in quite these terms. It fits comfortably with widely used welfare schemes such as the Five Freedoms, the 12 principles and the Five Domains, and even lends itself to concepts such as quality of life and positive welfare. In fact, as we will see in more detail in the following chapters, it does not just fit in with such ideas; it underpins them, validates them, gives them concrete reality and even clears up some of the ambiguity that they often come with. Outside science, too, 'health and what animals want' is readily understood by everyone from pet owners to farmers to politicians and scientists. It gives people a chance to see what evidence about welfare actually looks like and how necessary that evidence is to making sure we have really improved the welfare of animals.

Neither component is sufficient on its own. Health without what animal want misses a key part of what many people mean by welfare (Dawkins 2008), while, as we shall see next, what animals want on its own does not ensure good welfare either (Fraser and Nicol 2011; Franks 2019). Welfare assessment needs both, not just one or the other. Together, they form a powerful partnership for defining animal welfare. But the very fact that each partner brings a different perspective itself raises the issue of how those different perspectives can be resolved when they point in different directions. Before we get on to showing just how powerful the combination of 'health' and 'what animals want' can be together, we have to understand what happens when they conflict.

3

Why Do Animals Want What Is Not Good for Them?

'Health and what animals want' provides a definition of animal welfare that is succinct, practical and easy to understand but it has an obvious problem: animals do not always want what is good for their physical health. This is true even in wild animals living in their natural environments. The fly caught in a Venus flytrap because it wanted the sugary bait, the fish lured to its death because it wanted the fake worm dangled in front of it by an angler fish and the zebra that wants to drink but then gets eaten by the crocodile waiting in the waterhole—all show that what animals want is not always good for them in the long run.

In captive and domesticated animals, we see even more examples of animals choosing to do things that are not good for their heath—eating too much food, plucking out their own feathers, damaging themselves through repeatedly performing stereotyped behaviour and so on. In this chapter, we will look at some of the reasons that this happens, but why, despite the fact that they sometimes come into direct conflict, the combination of health and what animals want is still the best way of defining what good welfare is.

Time lags and arms races

Evolution by natural selection does not result in animals being perfectly adapted to their environments (R. Dawkins 1982). If they were, none of them would ever get killed, predators would always be successful in their hunting and none of us would ever succumb to infectious diseases. Evolution leads not to perfection but to animals that get by because they manage to survive and reproduce more successfully than others. They have evolved in ways that keep them alive most, but not all, of the time.

Even this degree of success, however, is temporary and precarious because other organisms—predators or prey or the diseases that infect them—are constantly evolving too, so that what was good enough for one generation may not be good enough by the time their children grow up. Evolution means that organisms are in a constant state of catch-up, always under pressure to do better and never quite achieving perfection because the goal posts are always moving (Ridley 1993). The evolution of antibiotic

The Science of Animal Welfare: Understanding What Animals Want. Marian Stamp Dawkins, Oxford University Press (2021).
 DOI: 10.1093/oso/9780198848981.003.0003

resistance is one familiar example of how success (in this case, our ability to kill bacteria) can be destroyed in a short period of time by the microbes themselves undergoing evolutionary change and becoming resistant to our previously successful drugs (Hudson et al. 2017; Aldara-Kane et al. 2018). Change is always happening and animals are always adapted to a past environment, the one in which their ancestors grew up and survived. In this sense, most animals are evolutionarily out of date.

One of the consequences of being out of date is that what animals were adapted to want in the past may not be what is best for their health and survival in the present. A poignant example of this happening to a wild animal is what happened to hatchling marine turtles when there was a relatively small change in their environment. When female turtles come ashore to lay their eggs, they choose nesting sites that are far enough up the beach that the eggs do not get drowned by an incoming tide. That means that when the young turtles hatch, they have to find their way back down the beach to the sea in which they will spend the rest of their lives (Figure 3.1). Loggerhead turtle (*Caretta caretta*) hatchlings recognize 'sea' as a strip of light with the brightest, lowest horizon (Kawamura et al. 2009; Limpus and Kamrowski 2013). For millions of years, this simple recognition rule worked perfectly well because a moonlit sea was the brightest, lowest source of light around. But then humans started erecting buildings with bright lights on the side of beaches away from sea. To the turtle hatchlings, the lights looked like a particularly inviting, shiny sea so they started to move towards the bright lights, directly away from the sea and often with fatal results. A previously adaptive tendency to want to go towards a strip of bright lights had become, in a changed environment, a threat to their health and very survival.

An even more widespread conflict between wants and long-term health is seen in what has become known as the global obesity epidemic in humans (Kopelman 2000) and is also recognized as a major health issue in companion, zoo, laboratory and farm animals (German 2006; D'Eath et al. 2009). Although there are many factors that contribute to obesity (Hill and Peters 1998; Rolls 2007; Guyenet and Schwartz 2012), one of the main ones is that, when faced with abundant food, humans and other animals want to eat more than is good for their long-term health. This again appears to be an out-of-date response to a new environment. In our evolutionary past, before there was abundant, calorie-rich food on supermarket shelves, before there was even agriculture, our hominid ancestors obtained their food by hunting and foraging. Their food was subject to massive fluctuations in availability in which periods of abundant food were followed by times of severe shortage. Our modern eating habits regarding what and how much to eat have been described as adapted to meet these periods of food shortage, when individuals who consumed or even overconsumed food when it was available were the ones most likely to survive the next time of hardship (Illius et al. 2002; Lieberman 2006). There was heavy selection to evolve physiological and behavioural mechanisms for guarding against weight loss in the face of a lack of food but less selection to avoid weight gain when faced with an abundance of food (Hill and Peters 1998) for the simple reason that in the environment in which our ancestors lived, death by starvation was a greater threat than death by overindulgence. Now, for both humans and other domestic species or those we keep in captivity, the previously successful strategy of wanting to eat

Figure 3.1 *An example of how what is best for long-term health and what an animal wants can become separated. To survive, newly hatched baby turtles have to find their way down the beach to the sea and they achieve this by 'wanting' to go towards the brightest strip of light around (normally the sea). But human buildings provide light that they want to approach even more, even though this leads them away from the sea and often to their death. Photo credit: noga f (Shutterstock).*

food when it is available is no longer aligned with what is best for long-term health (Illius et al. 2002; Lieberman 2006; Rolls 2012).

In addition, food manufacturers have deliberately designed food to be more palatable and tempting than would have been the case during the evolution of our feeding control systems (Rolls 2012). They have exploited our liking for sweet and fatty foods and turned them into 'supernormal' pleasurable stimuli such as ice-cream and chocolate which we find so delicious that they override the mechanisms that are supposed to stop us eating, such as gastric distension and glucose utilization (Bray 2004). As a result, we want to keep eating, and go on to eat more than is good for us. This has helped to fuel the rise in diseases associated with being overweight, such as diabetes, reproductive disorders, joint problems, cardiovascular disease and certain types of cancer (Hill and Peters 1998; Schwartz and Porte 2005).

Figure 3.2 *An obese cat is a living demonstration of what can happen when what an animal wants comes into conflict with what is good for its health. Domestic cats and dogs often want more food than is good for their long-term health. Wanting any food available is a good strategy when food is scarce but leads to poor health in a domestic environment with abundant food. Photo credit: Veda J Gonzalez (Shutterstock).*

Much the same has happened with domestic animals such as dogs and cats, which are often fed predictable, energy-rich food and required to make far less effort to get it than their wild ancestors would have had to make. Wild carnivores such as wolves and cheetahs expend considerable time and effort in chasing and bringing down their food with as many as half of their chases being unsuccessful and resulting in nothing to eat at all (Sand et al. 2006; Hilborn et al. 2012). With food being so difficult to come by, they have to eat when the opportunity presents itself. Domestic dogs and cats following the same strategy of eating when they can but given too much food and not enough exercise tend to become overweight and to develop similar health problems to obese humans (German 2006; Courcier et al. 2010) (Figure 3.2).

It is quite clear, therefore, that we should never rely solely on what animals want to assess their welfare because what they want when left to themselves is not necessarily what is best for their long-term health any more than it is with us humans. This is particularly true for animals now kept in environments that are very different from the ones in which their ancestors evolved and where the responses that were once adaptive may no longer be so. Such animals are found on farms, in zoos, in laboratories and in people's homes—the very places where concern for animal welfare is greatest (Gygax and

Hillman 2018). Giving these animals everything they want without taking into account what is best for their long-term health will not ensure good welfare. At the same time, considering only what makes the animals physically healthy in the long term without also taking into account what they themselves want in the here and now would, as already discussed in Chapter 2, leaves us with an inadequate account of what 'good welfare' means. That is why we need both as equal partners in the definition. What animals want must always be taken together with what is good for their health and vice versa.

Balancing health and what animals want

As the obesity example shows, a conflict between what we want and what is good for us is not unique to animal welfare. We take children to the dentist or to be vaccinated even though they don't want to go because we have decided it is good for them in the long run. We pass laws that compel people to wear seat belts or comply with health and safety rules, overriding what they might want to do if left to themselves in favour of what is good for them and others in the long run. We don't just let people choose to smoke all the cigarettes or take all the drugs they want but try to stop them out of concern for their long-term health. In fact, as I write, countries around the world are considering legislation to try and alter what food people want to eat, by putting taxes on sugary foods that their governments have decided is bad for their long-term health. The legislation is considered necessary because people want these foods so much that it may be the only way to override the power of their own personal choices.

So, in human affairs, we acknowledge that conflicts can exist between short-term wants and long-term health needs, and we do not have a single solution for all such human dilemmas or any right way of deciding how much weight should be given to one or the other. In fact, there is a large measure of disagreement about the balance that should exist between the 'nanny state' that decides what is best for people's long-term interests and their personal freedom to choose for themselves, but this is the realm of ethics not science, of what ought to be rather than the facts as they are. Similarly, with animal welfare, there is no simple formula for deciding how to balance what animals want with what is best for their health when these two come into conflict. As with comparable human dilemmas, there are many different opinions as to which one matters most. This book is not intended to make ethical judgements but simply to equip you to make up your own mind in the light of what the facts are. The facts that are important are what animals want and what is best for their long-term health. The conflict between the two, where it exists, needs to be pointed out, so that everyone can weigh them up to make their own judgement.

Why animals generally do want what is good for them

Having just looked at why what animals want is not always good for their health, it must also be emphasized that, most of the time, these two measures are very closely aligned and the conflict between them is much less than might be expected. Animals may sometimes make 'bad' choices, especially when they are in new or unusual environments, but what they want can often give a valuable indication of what might be beneficial for their health. For example, chickens choose to eat small stones and grit because they don't have teeth and stones are essential for their ability to break down and digest food in the gizzard. Giving them what they want in this instance improves their digestion and therefore their health.

Another even more obvious example is that, when they are very young, baby chicks are unable to regulate their own body temperature and need external sources of heat, such as their mother's body or artificial heat lamps. They also want heat and will utter high-pitched piercing 'distress' calls if they are too cold (Wood-Gush 1971). Chicks kept at the temperature they want are healthier and thrive better than those that are kept colder than they want (Dawkins et al. 2004), so ignoring what the chicks want in this respect results in birds with poorer health.

The generally close connection between health and what animals want exists because wanting to obtain the right things and wanting to avoid the wrong ones are major ways in which animals keep themselves healthy. Animals have evolved many different ways of maintaining their health and then regaining it again once it has been compromised, such as an ability to heal wounds when they are injured and an amazingly complex immune system for warding off infection. Animals are equally good, however, at dealing with injury and disease before they even happen. They have evolved a complex set of mechanisms for anticipating and avoiding danger altogether. They can take pre-emptive action so that the worst never happens. They start to want things that will be necessary for their health and survival not for now but for some time in the future.

We can see how this remarkable, future-predicting ability operates with two examples. Animals eat food because they need fuel to keep them alive and healthy, but they frequently start wanting and looking for food long before their bodies are in real danger of running out of food reserves altogether. This early activation of behaviour associated with wanting food (which we call hunger) reduces the risk of future starvation rather in the way that the fuel gauge on a car warns of future, not immediate, absence of fuel. Animals thus have a long-term need for food (without which their health would deteriorate) but they also start to want food before this need becomes urgent. Satisfying this want early protects their health in the long term.

The way in which animals can appear to anticipate future threats to their health and survival can also be seen in animals that live in social groups. Many animals group together as protection against predators (Krause and Ruxton 2002) and have evolved mechanisms for recognizing and remaining close to other members of their species. They want to stay with other animals even when there are no predators in sight because one day predators might appear and then the protection of the group will become essential to avoiding injury and death. An animal that wants to be with other members of its species even when there are no predators around is thus preparing for a dangerous future that has not yet happened, but which is very likely to. This wanting to be with other animals can be so strong that it takes on a life of its own and becomes a powerful driver of an animal's behaviour even when the animal is completely protected from predation, for example, in a zoo or farm. An animal can become very agitated and make great efforts to find companions even when it is quite safe because its ancestors were selected to 'want' to be in a group and, as a result, survived.

In these two examples, starvation or being injured by a predator would constitute one part of 'poor welfare' as defined by an immediate decline in health status, but vainly searching for food or companions would constitute the second component of 'poor welfare' as defined by the animals wanting something they do not have. The two components are closely linked. The animals may be physically healthy now but if death in the form of a predator or not getting enough to eat is just around the corner, then those possible deaths can be avoided by the animals taking pre-emptive action of obtaining food or companions immediately. Wild animals have, in general, evolved to want things in the short term that serve their physical health and fitness in the long term. Problems arise when that link is broken.

Health and what animals want

Many of the most serious animal welfare problems that confront us are the direct consequence of animals *not* having what they want by repeatedly trying to get what they want and persistently failing. Sometimes, persistent, unfulfilled wanting can lead to behaviour in which animals try so hard to get what they want that they physically injure themselves. For example, they might throw themselves against the bars of a cage in an effort to escape. At other times, animals such as caged migratory birds and confined tigers and polar bears may look healthy but this can hide long-term physiological, immunological and behavioural effects of being repeatedly unable to do what they want (Clubb and Mason 2003; Vaz et al. 2017). If this continues for long enough, the results can become pathological in the form of stereotypies or actual injury, as we will discuss in more detail in Chapter 5.

In humans, it is increasingly acknowledged that there is a link between mental health and physical health. For example, immune function is highest in people who report being happiest and have what they want in life (Nakata et al. 2010; Takao et al. 2018). Conversely, impaired immune function has been found in people deprived of basics such as not having a home (Arranz et al. 2009). What animals want gives us, despite the problems and reservations we have already noted, valuable information about what they might need to keep them healthy in the long term. Not infallible information, certainly, and not information that can be unquestioningly used on its own, but valuable pointers that could provide important new information.

For example, giving animals what they want most just might be the best vaccine we could ever have for stress-free, disease-free animals that do not require medication (Dawkins 2019). Although there is no simple relationship between immunity and welfare (Boissy et al. 2007; Berghman 2016), reduced immune function and greater susceptibility to disease are widely recognized as results of poor welfare (Gross and Siegel 1981; Moberg 1985; Broom and Johnson 1993; Cockram and Hughes 2011). In humans, good immune function is closely related to peoples' subjective reports of being happy and satisfied with their lives (having what they want, in other words) (Nakata et al 2010;

Takao et al. 2018). At the very least seeing how giving non-human animals what they want impacts on their long-term health could give us new ideas for better management and treatment that we have hardly begun to explore. What animals want is thus a key part of what we mean by welfare both for what it tells us about the animals themselves and for the guidance it can potentially provide about what they need to keep them healthy.

Conclusions

So far, we have seen that there is more to good welfare than just keeping animals physically healthy and that 'what animals want' supplies much of what is missing from a definition that relies solely on health. We have also seen that animals do not always choose what is good for their long-term health, so good welfare needs to include both what leads to physical health and what animals themselves want. There are evolutionary reasons why these two elements of welfare sometimes conflict and even stronger reasons why they generally reinforce each other. Taken together, these two elements form a team, a comprehensive and comprehensible definition of welfare and one that is, most importantly, firmly rooted in what we can actually demonstrate about an animal's behaviour and physiology.

We now turn to the crucial question of how we can discover what it is that animals want. The answer lies in finding the right ways to 'ask' them.

4

What Animals Want

'Asking' animals what they want when we cannot use words might seem far-fetched and even fanciful, but one of the great achievements of animal welfare science is to have developed a variety of different ways of finding objective and scientific ways of doing just that. It turns out that many of the methods that are used are quite similar to the ways we determine what other people want because even with other people, of course, words are not everything. 'Actions speak louder than words', we say or 'He put his money where his mouth is', implying that we have more trust in what people do than in what they say. And 'They voted with their feet' is a far more convincing statement of what people truly want than any number of words they might use.

So it is with other species. Animals may not have language as we know it but they do have many different ways of obtaining what they want and for getting away from what they do not want. What we have to do is to tap into these different actions and then interpret them correctly.

One-off choice tests

The simplest method of establishing how an animal views something is to present it with a choice between two or more alternatives, and see which one it chooses. Hughes and Black (1973) asked what sort of flooring hens wanted by offering them a choice between two different types of floor and then measuring which floor they chose to stand on for the longest amount of time. As it turned out, the hens did have a definite preference but, somewhat controversially, the floor they preferred was not the one that had recently been recommended by a committee set up by the UK government to look into the welfare of intensively housed farm animals. The choice in front of the hens was between two different kinds of wire floor—fine-gauge wire or heavy rectangular mesh—both at that time used in battery cages that were then the commonest way of keeping egg-laying hens (battery cages have, incidentally, since been banned by the EU and in many other places around the world). When given the choice, the hens spent more time standing on the fine wire than on the heavy rectangular mesh that had been recommended by the committee (Brambell 1965). With the best of intentions, the committee had recommended the use of heavy rectangular mesh on the grounds that they thought it would be more comfortable for the hens' feet. The hens' own choices showed otherwise.

The Science of Animal Welfare: Understanding What Animals Want. Marian Stamp Dawkins, Oxford University Press (2021).
© Marian Stamp Dawkins. DOI: 10.1093/oso/9780198848981.003.0004

This example shows up both the strengths and weaknesses of a simple one-off choice method. On the one hand, it allows animals to directly compare alternatives and to deliver their verdict in terms of what they choose to do, even when, as in this case, it goes against well-meaning human assumptions. On the other hand, they can only express their preference among a very limited range of options—the ones that humans provide for them—and it may be that neither fine-gauge chicken wire nor heavy-gauge metal mesh is particularly comfortable for chickens' feet. The chickens may simply have been opting for the flooring they disliked least (Duncan 1992). Alternatively, they might have chosen the fine-gauge wire because that was the only type of flooring they were familiar with and they were wary of anything new.

For these and other reasons, there has been a gradual refinement in how choice tests are carried out, with thought given to what options should be offered and also to what animals have experienced previously (Fraser and Nicol 2011). It is now acknowledged that it is important to look not just at what animals choose when first offered a choice but also how they respond to each option when they have had more experience of it. To use the terminology we met in Chapter 2, we need to know not just what animals want, but also whether they like it when they get what they want (Berridge et al. 2009), at least enough to choose it again. Most choice tests are now run after making sure that animals are equally familiar with all the options being offered to them so that they don't just choose what they have known before and can make informed choices based on their previous experience with each option. What they choose thus becomes a true reflection of what they both want and like.

Repeated choice tests

As the hen flooring experiment shows, giving animals single choices between a limited set of alternatives has its problems (Fraser and Matthews 1997; Fraser and Nicol 2011) and an obvious improvement is to widen the range of options available, thus overcoming the problem of an animal having to choose what it dislikes least. A second improvement is to offer the same choice over and over again, so that an animal gradually builds up experience of what it is choosing and so can indicate whether that experience makes it more or less likely to choose that option again. Its choices become informed choices.

For example, to establish a horse's point of view of being ridden in a certain dressage posture, von Borstel et al. (2009) offered horses a choice between being ridden with their heads in a normal position and being ridden with their heads forced into an unnatural dressage position called Rollkur in which its neck is hyperflexed by being pulled towards the chest (Figure 4.1). They used a large Y-maze to offer the choice and gave the horses experience of each of the two riding positions by riding them repeatedly through the Y-maze (Figure 4.2).

At the end of one arm, the rider held the horse's neck in the Rollkur position and then rode the horse in a 20-m circle in this position either walking or at a trot. At the end of the other arm of the Y-maze, the horse was ridden in the same circle and at the same two paces but with its neck in a normal posture. The horses were trained for 15 trials with the

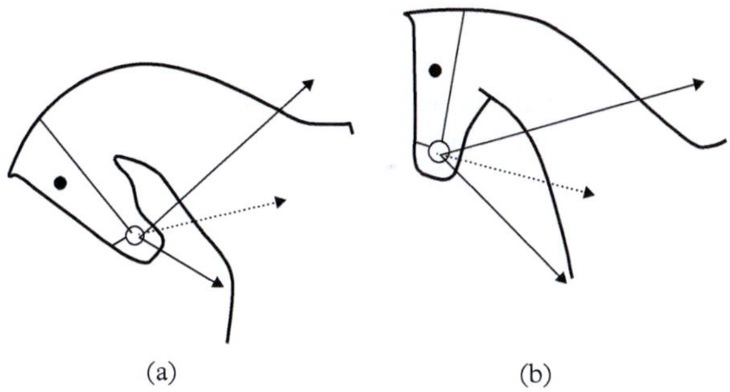

Figure 4.1 *Horse in (a) Rollkur dressage position and (b) normal riding position. The dressage position of Rollkur involves the rider applying pressure through the reins (top arrow) so that the horse's muzzle almost touches its chest. Rollkur makes it difficult for the horse to see in the direction in which it is travelling and potentially disturbs its balance. Reprinted from Von Borstel et al. 2009 with permission from Elsevier.*

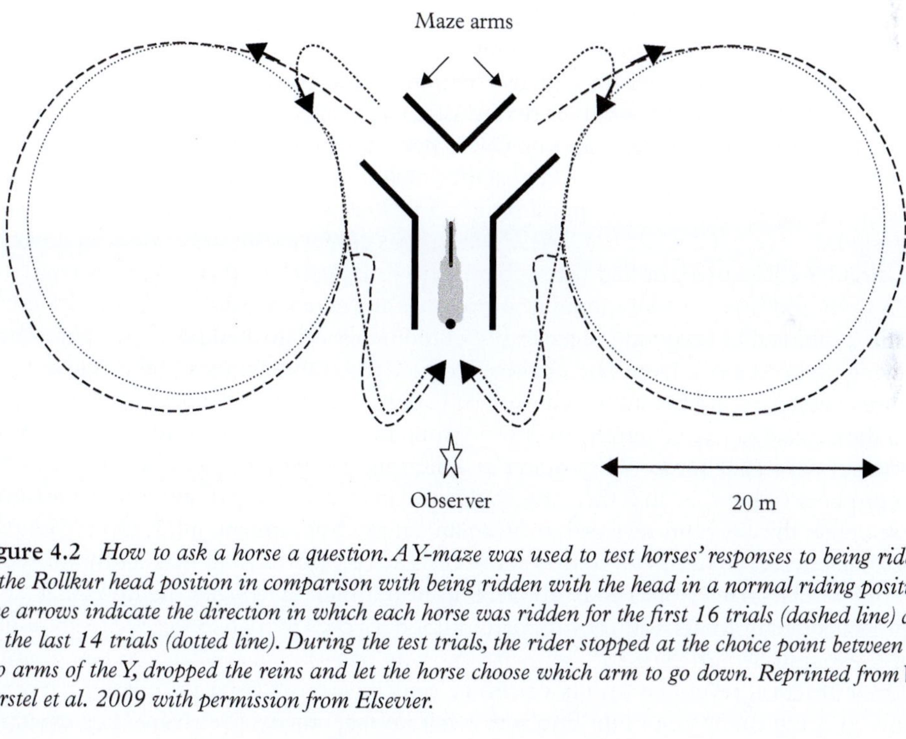

Figure 4.2 *How to ask a horse a question. A Y-maze was used to test horses' responses to being ridden in the Rollkur head position in comparison with being ridden with the head in a normal riding position. The arrows indicate the direction in which each horse was ridden for the first 16 trials (dashed line) and for the last 14 trials (dotted line). During the test trials, the rider stopped at the choice point between the two arms of the Y, dropped the reins and let the horse choose which arm to go down. Reprinted from Von Borstel et al. 2009 with permission from Elsevier.*

Rollkur treatment and 15 trials with the normal treatment given in pseudo-random order (so that they had experienced each treatment equally recently). After this training to make sure the horses knew what they were in for, the choice tests began. The rider rode the horse into the Y-maze but then stopped at the choice point, dropped the reins and allowed the horse to choose which arm to go down. A total of 14 out of 15 horses consistently chose the arm that resulted in them being ridden with the normal posture (von Borstel et al. 2009). Most of them were very clear that they did not want to have their heads forced into the Rollkur position.

In a somewhat similar experiment to see what sort of human contact cows preferred, Pajor et al. (2003) also used a Y-maze and offered cows a choice between two sorts of treatment by humans. If they chose one arm of the Y, the cows met a person who spoke softly to them. If they chose the other arm, they met someone who shouted 'Hey, you stupid cow!' in a loud voice. Each cow was exposed to each option, but with just two trials of each experience the cows reliably chose the arm where they were treated with a gentle voice (Pajor et al. 2003). The cows, too, were clear what they wanted.

Repeated choice tests have proved a very useful way of understanding what animals want because they do not rely on an animal's first response to a choice, but on what it chooses when it has experience of what it is choosing. An animal that makes the same choice repeatedly must 'like' what it chooses because it knows what it is in for. Experience of the choices on offer does not, however, have to come through formal trials as in the horse and cow examples above, but it can also be achieved through letting animals explore the choices for themselves. The important thing is to be sure that the animals have experienced all the options equally when they make their choices.

For example, to find out whether hens dislike the smell of ammonia, Kristensen et al. (2000) offered them a three-way choice between fresh air and two concentrations of ammonia—25 and 44 ppm. To humans, the smell of ammonia is very unpleasant even at low concentrations and an upper limit of 25 ppm is widely imposed for human comfort and safety (e.g. CDC 2012). The choice in this case was offered to hens in the form of a circular enclosure that had three chambers—one with fresh air, one with ammonia at 25 ppm and one with ammonia at 44 ppm. The three chambers were separated by plastic curtains that retained the gases but could be easily pushed through by the hens, allowing them to move from one chamber to the next. Only when all hens had entered all three chambers did the actual choice test begin. Once the hens knew what was available, they chose to spend significantly more time in the fresh air chamber than in either of the ammonia chambers. They spent as little time in the 25 ppm chamber as in the 44 ppm one, suggesting that their threshold for choosing to avoid ammonia was something below the 25 ppm level set for human safety. Subsequent work showed that, if given the choice, chickens avoid ammonia even at 10 ppm, a level frequently exceeded on commercial farms (Jones et al. 2005). Pigs also dislike the smell of ammonia at similarly low concentrations and they too avoid it where they can (Jones et al. 1996; Smith et al. 1996; Nielsen 2018).

On a practical level, the results of choice tests have been used to redesign environments to bring them more into line with what animals themselves want. For example, old-fashioned mouse cages that were previously little more than bare plastic boxes now routinely include nestboxes, platforms, pipes and other enrichments, and this is at least partly on the grounds that the animals themselves have shown a preference for cages with these enrichments (Bradshaw and Poling 1991; Townsend 1997; Van de Weerd et al. 1998; Smith and Corrow 2005; Baumans and Van Loo 2013). The preference of green sea turtles (*Chelonia mydas*) for shade over sun has been an important factor in designing ways of farming them successfully (Keller and Mustin 2017). Chickens offered a choice between roosting sites that were 60 cm above the ground and those that were only 15 cm above showed a clear preference for roosting at night in the higher places, even if there were no proper perches there (Schrader and Mueller 2009). Experienced birds clearly want to roost high up—giving us another indication of how to design animal-centred housing.

The choices of the animals themselves have even contributed to the debate about the most humane way of killing rats and mice. Rats choose to leave a chamber containing either CO_2 or argon (both frequently used for euthanasia) even when they are in much lower concentrations than are lethal to them (Niel and Weary 2007; Kirkden et al. 2008). Other gases such as sevoflurane rats find less aversive (Guedes et al. 2017).

One of the oddest examples of animals showing what they want is that of wild rodents such as wood mice choosing to re-enter traps in which they have previously been confined (Hernandez et al. 2018). This appears to be partly because the mice learn that there is food inside the trap but also because the trap offers protection against predators. Mice were found to be more likely to enter traps when fox faeces were scattered in the vicinity (Hernandez et al. 2018). Even more bizarrely, wild mice, shrews, rats and frogs have all been recorded on video choosing to run in a running wheel—the sort that is sold for mouse cages—when one was left outside. Despite the fact that they were completely free to run anywhere else they liked, the wild mice entered the running wheel and started running as vigorously as any pet mouse (Meijer and Robbers 2014). We clearly need to keep an open mind about what it is that animals want!

Working for what is wanted: operant conditioning

Giving animals the opportunity to make choices is obviously a powerful way of demonstrating what they want or do not want. But many of the choices we have discussed so far are simply what animals would do anyway if they want something—pushing past a barrier, moving into the shade, turning one way or another. These could all be hard-wired, innate responses, done automatically and unthinkingly, like blinking or breathing. Particular interest in animal welfare science has therefore centred on going beyond these innate responses and finding ways of showing that animals have more complex and cognitive ways of getting what they want. A large body of evidence now shows that all vertebrates and some invertebrates do indeed have the flexibility of behaviour to work out how to get what they want when their innate behaviour fails to achieve their goals and they have to learn to do something completely different. This has opened up a whole new way of studying what animals want.

If an animal can be trained to perform some novel behaviour thought up by a human, such as pressing a lever on a machine, swimming through a hoop or performing a trick, just for the result of obtaining the reward of a piece of food, then that animal has given us two important pieces of information. First, it has told us that it wants that food enough to be prepared to do highly unnatural behaviour to get it. Second, it has shown that it must possesses a mechanism for getting what it wants that is a step change from a simple innate hard-wired response of approach or avoidance. If the animal can then be retrained so that it has to do the opposite of what it initially learnt such as pushing the lever upwards or swimming the other way through the hoop or learning a completely different trick—and it relearns the task to get the reward—then it is even clearer that the animal has the goal of obtaining food and is capable finding many different ways of achieving that goal. This process in which animals can learn entirely new behaviour to get what they want is called operant conditioning (sometimes referred to as reinforcement learning) and it gives us a direct window into what animals want (Kilgour et al. 1991; Patterson-Kane et al. 2008).

This window exists because the ability to learn new behaviour has evolved as one of the main mechanisms by which animals can obtain what they want in nature, giving them far more flexibility than if all they could do was follow fixed innate rules. For example, many different foods like honey and fruit have a sweet taste but all look very different and require different ways of obtaining them. By having an innate liking for a sweet taste (finding a sweet taste 'rewarding' in other words) and combining this with an ability to learn many new behaviours to give themselves that sweet taste, animals can achieve far more flexibility and adaptability to a changing unpredictable world than if all their behaviour were innate and fixed in advance. They can discover all sorts of new sources of food never encountered by their ancestors through a subtle but constructive interplay between genes and environment. Their genes specify what they find rewarding or punishing, guiding them towards what is nutritious and away from what is harmful. But then they learn from their own experience what to do to achieve those goals or rewards (Balleine and Dickinson 1998; Dickinson 2012; Pennartz et al. 2019). So by using operant conditioning to find out what animals find rewarding or punishing, we humans have a powerful way of accessing what matters to them. Operant conditioning can be seen as giving all the information we could get from repeated choice tests, but with the additional bonus that if animals will learn do something completely unnatural to bring about a particular result, we have really understood what is guiding and controlling their behaviour (Figure 4.3). We have understood what they want and what they like.

It is perhaps not surprising that many animals will learn to perform operant responses for a food reward since obtaining enough food is vital to the survival of all species, but animals will also learn tasks for a variety of different rewards. Mice will learn to press a switch for the reward of bedding material (Roper 1976) and also for being allowed access to damp moss-peat that they can burrow in (Sherwin et al. 2004). Blue foxes (*Vulpes lagopus*) easily learn to press a lever for the reward of being allowed somewhere to dig (Koistinen et al. 2007), hens will learn to peck a key to gain access to straw (Gunnarsson et al. 2000), and pigs, calves and sheep—at least those that have been shorn—will learn to operate switches for the reward of heat (Baldwin and Meese 1977; Baldwin and Start 1981; Jones and Nicol 1998). Heifers have learnt to operate a panel with their noses to give themselves 15 minutes of being able to lie down (Jensen et al. 2004) and calves will learn to push a panel with their heads for the reward of a few minutes' social contact with another familiar calf (Holm et al. 2002).

Figure 4.3 *How the combination of genetics (an innate liking for certain tastes, which constitutes a 'reward') and learning (the ability to link a wide range of behaviours to the subsequent obtaining of that reward) results in highly adaptive, flexible behaviour. Grey squirrels (Sciurus carolinensis) are adept at obtaining food from bird feeders by learning to perform highly unusual behaviour to get the reward that they want. Photo credit: Joe Kravetz (Shutterstock).*

An interesting addition to this list of what constitutes a reward is that some animals even find the operant conditioning tests themselves so rewarding that they will choose to use the operant equipment to get some of their food and water even when they can get the same rewards 'for free' from a normal food or water dish. When goats had to learn to press a button for a reward of 35 ml of water, they still chose to obtain some of their water by operating the machine and getting only this small amount each time, even though there was plenty of drinking water freely available just nearby and their rate of obtaining water from the machine was much lower than drinking in a more natural way (Langbein et al. 2009). In fact, animals of many different species seem to find it so rewarding to 'work' for their food by performing operant responses that it has even been suggested that operant conditioning equipment might be regularly installed as an 'enrichment' for zoo animals (Meehan and Mench 2007; Westland 2014).

Choice tests for the wider world: bigger questions and bigger choices

The choice and operant conditioning tests we have discussed so far show how animals can be asked for their views on very specific aspects of the way in which humans treat them, such as how they interact with them and what sorts of living quarters they prefer. This has led to the view that such tests have only limited value because they cannot provide answers to the really big issues in animal welfare, such as whether farm animals should be kept outside and allowed to range freely, whether tourists adversely affect the welfare of wild animals or whether zoos can ever provide conditions that ensure the welfare of wild animals. But, properly designed, choice tests can be extended to answer a very wide range of questions.

For example, dairy cows are increasingly kept indoors for part or all of the year and fed concentrated food to increase milk yield. Human consumers have a strong view that this is not right and that cows should be able to go outside and eat grass (Cardoso et al. 2018). It is therefore of considerable public interest to know whether the cows themselves want to go outside. Charlton et al. (2011a) gave dairy cows a choice between indoor housing where they could obtain high-energy food and going out to pasture where they could graze freely. Twice a day after milking, they took individual cows to a choice point equidistant (48 m) from indoor housing and pasture and let them choose where to go. Between milkings, the cows could move between the two areas so that they had ample experience of both. When taken to the choice point, the cows were more likely to choose the pasture than the indoor stalls (66.2 per cent v. 33.8 per cent) but overall spent more time indoors, although the behaviour was highly dependent on the weather, the individual cow, the time of year and the time of day (Falk et al 2012; Charlton and Rutter 2017). Cows do, however, show a strong preference for lying down outside at night (Charlton et al. 2011b; Lee et al. 2013; Charlton and Rutter 2017; Smid et al. 2018).

Such grand-scale experiments providing animals with choices between two very different environments are, however, logistically very difficult to set up and are often highly impractical. Fortunately, the concept of 'choice' can be extended way beyond specially constructed Y-mazes or runways by borrowing some of the methods used for studying wild animals. Ecologists frequently talk about the habitat choices of wild animals such as leopards (Balme et al. 2007) and foxes (Van Etten et al. 2007), or how birds choose particular nest sites (e.g. Magana et al. 2010) even though they never actually offer them any sort of formal choice at all. They simply document where an animal is found or where it builds its nest and use this as what it has 'chosen', relying on the implication that, because it is a wild animal, it could have chosen anywhere else. 'The rest of the world' thus becomes the non-chosen option and where the animal is actually found is taken as its preferred option. Of course this has to be modified because even wild animals do not have entirely free choices (another animal may be occupying their preferred territory, for example), but nevertheless where a wild animal chooses to live when it has a range of other options open to it is an important clue as to what it wants and is one of the reasons that natural behaviour has assumed such importance in animal welfare (Chapter 6). Tracking devices that continuously record where animals go and what they do even when they are out of sight of humans have opened up a whole new world of understanding what wild animals want as measured by where they go, what they avoid and where they spend their time.

This ecological approach to choice of using where animals choose to go is being increasingly used to document the choices of domesticated and zoo animals, particularly those in large enclosures with diverse areas where the animals have a genuine choice of where to go. For example, zoos can find out whether animals find the presence of visitors aversive or attractive by providing refuges where the animals can hide from visitors and then seeing whether the animal choose to move close to visitors or to move away from them (Bloomfield et al. 2015; Troxell-Smith et al. 2017; Brando and Buchanon-Smith 2018). It is also possible to establish what physical features of their enclosures animals like or dislike by recording where they spend their time (Ross et al. 2009). In an agricultural context, both free-range broiler chickens and free-range laying hens spend a disproportionate amount of their time either under or near trees if this option is available to them (Dawkins et al. 2003; Larsen et al. 2017), a clear indication of 'choice' and that the cover of trees is what these animals want.

Social preferences—that is, whether animals want to approach or avoid other animals—can also be studied by looking at how they position themselves with respect to each other, particularly the distances between them (McBride et al. 1963; Stricklin et al. 1979; Keeling and Duncan 1989; 1991; Keeling 1995). Thus, animals that want to be in close proximity to other animals will show this by arranging themselves into groups or clusters, whereas those that want to avoid each other will attempt to get as far away as possible from each other, use up all the space available and space themselves regularly. Sometimes, as in nesting colonies of seabirds, animals will try to do both at the same time. Black-headed gulls (*Larus ridibundus*), for example, cluster together and build their nests in a relatively small area as a protection against predators, but, within the colony, the nests are regularly spaced and never less than about 1 m apart, kept at that distance by aggression between the nesting pairs (Kruuk 1964; Patterson 1965). The actual spacing of the nests (clustered into colonies but maximally spaced within a colony) reflects the gulls' conflict between wanting to be together for protection but at the same time wanting not to be too close to prevent neighbours cannibalizing their eggs and young. Figure 4.4 shows a dramatic example of such a compromise in another colonially nesting seabird, the northern gannet (*Morus bassana*), where the regular spacing of the nests results in the birds nesting in uniform rows.

Spacing patterns are particularly useful for looking at what animals want when actual choice tests are not feasible or difficult to devise, for example, in investigating whether animals want more space or find high stocking densities aversive. Because they do not involve disturbing the animals in any way, they can be very sensitive and reveal subtleties in what animals want that other more intrusive methods can miss. When they are very young, for example, broiler chickens choose to cluster together and want to be together (Febrer et al. 2006), but when they are older they tend to space themselves out and find it aversive to be too close to other birds (Buijs et al. 2011a, b).

Figure 4.4 *What animals want revealed by their spacing patterns. Many seabirds, including these northern gannets (*Morus bassanus*) nest in dense colonies as protection against predators, but, within colonies, space out and keep their distance from each other. The result, as here, is near uniform spacing of nests in almost straight lines. The observed spacing pattern reflects a compromise between the birds wanting to be close to each other, but also not wanting to be too close.*

Asking how much animals want something

To relate 'choice' to 'welfare', we need to know not just what an animal wants but how much it wants it. An animal might have a clear preference for one option over another but if that second option is something it wants almost as much as the first, it would be difficult to argue that its welfare was compromised if only the second choice were available. It would be like us being offered a choice of our two favourite foods and having to make do with just one of them. We would be spoilt for choice but well off whichever we chose. On the other hand, if we had to choose between two foods we really hated then having to eat the least bad option would not necessarily be pleasant or what we really wanted, even though we might have (reluctantly) have chosen it. In other words, choice just gives us a ranking between options. It does not tell us whether animals want and like all of them or dislike all of them. Thus, to go from choosing to wanting, we need to know how much they want something in comparison to something we know they want, most obviously food (Dawkins 1983, 1990). We need to know, for example, if they want it so much that they will still choose it if they have to pay some sort of price for it and whether they will pay the same price as they would for food, rather in the way we might decide we would quite like something, but not enough to walk for a mile in the pouring rain to get it.

For animals, paying a price might be having to squeeze through a narrow gap (Sherwin and Nicol 1995; Bubier 1996), walk down a long corridor (Bokkers et al. 2007), push a weighted door (Mason et al. 2001; Olsson and Keeling 2005), walk through a water bath (Sherwin and Nicol 1996; Dixon et al. 2014), run over an electric grid (Walker and Mason 2018), negotiate an air blast (Faure and Lagadic 1994) or simply repeatedly perform an operant response such as pecking a key not just once but many times for each reward. The way the price is paid can be varied to suit the animal and the circumstances.

What is important, however, is that it must be possible to vary the cost an animal has to pay in a systematic way so that the animal can initially be offered a choice with little or no cost—such as pushing through a door with no weights or obtaining food after a single key peck—and then be offered that same choice when the price is made progressively higher—such as having to push a door with heavier weights or having to make many key pecks for just one reward. Exactly how the price paid for different commodities should be compared has been the subject of lively debate (e.g. Matthews and Ladewig 1994; Mason et al. 1998; Kirkden et al. 2003; Patterson-Kane et al. 2008), but one of the most widely used measures is the Maximum Price Paid (MPP), which is the price an animal is willing to pay in the face of increasing costs before it gives up and refuses to pay any more. By comparing the MPP for different rewards, including food, it is possible to directly compare how much animals want different things. An animal that will pay as much for, say, access to bedding material as it will pay for food is showing a real tangible want for bedding, not just a mild preference. For example, American mink (*Mustela vison*) showed that they wanted water to swim and were prepared to pay a considerable price for it. Mason et al. (2001) kept mink in a complex environment that had doors leading to different resources, such as a tunnel, a raised viewing platform and a swimming bath. The mink thus had a multi-way choice but had to push open a door to get what they wanted. When the doors were unweighted and easy to push, the mink used all the resources, but when the doors were made heavier, they stopped using some of them. For the water bath, however, they kept pushing even when the doors were weighed with 1.25 kg, about the maximum a mink can push. Mink very much want to be able to swim in water.

Another example of animals being willing to pay a high cost is to be found in the efforts that dairy cows are willing to make for access to automated rotating brushes, which they then rub against different parts of their bodies (Westerath et al. 2014; Figure 4.5). When they are outside, cows use trees or fence posts to clean themselves. When kept inside without access to these natural grooming aids, cows will quickly learn to operate switches to make the brushes rotate and they will push open gates to get access to the automated brushes. They will keep pushing even when the gates are made heavier and heavier. The MPP (in this case, the heaviest gate they will push before giving up) is the same as the MPP for food when they have been food deprived for 1.5 hours (McConnachie et al. 2018). This suggests that cows want access to the grooming brushes

Figure 4.5 *Animals will work for many rewards besides food. Here a cow is using a mechanical rotating brush. Cows find being 'groomed' by the brushes highly rewarding and will learn to perform operant tasks to turn the motors on. Photo credit: Pixeljoy (Shutterstock).*

as strongly as they want access to food, at least at this deprivation. In acknowledgement of this, plus the fact that automated brushes keep the cows cleaner, automated rotating brushes are now widely used on dairy farms in countries such as Denmark. The cows have shown that they want brushes and that they want them very much.

How much animals want something can also be measured without making them choose in an all-or-nothing way but simply by how quickly they approach objects or situations they have experienced before. Abeyesinghe et al. (2001) used an ingenious version of this method to show that hens want to avoid the vibration and heat that occurs during transport on a truck. Hens were trained to run down corridors of about 4 m to find food in a goal box. After the hens were reliably running to find their food reward, they were suddenly confined in the goal box with no food and subjected to the kind of environment they might experience during transport—the whole goal box was vibrated vertically at 2 Hz and heated to 40°C to mimic the conditions on a truck. This went on for 60 minutes, while control birds just experienced an empty goal box for the same length of time. On the next 10 trials, the hens' speed of running towards the goal box was carefully measured. The hens that had experienced the 'transport' (vibration plus heat) conditions ran significantly slower than the control hens. What they had just experienced clearly made them reluctant to approach a goal box that all their previous training had told them was a reliable source of food. This drop in how quickly they ran towards something they really wanted (food) showed that they did not want the experience of 'transport'.

Rushen (1986a) used a similar method to show that sheep find the controversial practice of electro-immobilization (in which the sheep are stopped from struggling while being sheared by having electric current passed through them) particularly aversive. He put sheep into a corridor, which they naturally ran down to escape from people at the start end. At the other end of the corridor, some of the running sheep were caught and physically restrained by a person (as would happen in normal shearing), others were caught and then electrically immobilized (which makes things much easier for the shearer) and others were not restrained at all. Once the sheep had had experience of what was going to happen to them at the end of the corridor, Rushen then reran all the sheep down the corridor and measured their speed of running. Sheep that had experienced the full restraint with electro-immobilization went more and more slowly as the trials progressed and eventually became reluctant to approach the end of the runway at all, suggesting that this procedure was something they wanted to avoid. Sheep that had simply been restrained without electro-immobilization moved faster, but the sheep that were never caught at the end of the runway ran fastest of all, suggesting that being restrained by a human is something sheep want to avoid if they can. However, they want to avoid electrical immobilization more than anything.

Just as running more slowly can be used to assess what animals find aversive and want to avoid, so running faster can be used to assess what they find attractive and want to interact with. This method was used to ask rats which sort of 'enrichment' object they preferred. As it is now a requirement in many countries that laboratory mice and rats should not be kept in barren cages but should be given enrichments, it is obviously important to know what sorts of objects the rodents themselves want and to be sure that what humans think constitutes an 'enrichment' really is so from a rat or mouse point of view. Hanmer et al. (2010) trained rats to run down a corridor to a goal box where they were allowed 5 seconds of interaction with a training object. Once the rats were reliably running to the goal box, the researchers removed the training object and substituted different potential 'enrichment objects' such as a plastic ball or a cardboard tube in the goal box and gave the rats three trials with this new object. They then measured the speed with which the rats ran down the runway. They found that the rats ran at different speeds depending on what objects they had previously experienced in the goal box. They ran fastest towards objects that they had previously shown they preferred in a simple two-way choice test. Running speed thus gave a similar picture of what the animals wanted to an actual choice test, but also gave a continuously varying indication of how much they wanted it.

'Informed' choice when repeated choices are difficult or impossible

The validity of both repeated choice tests and operant conditioning depends on animals having enough experience of the various options that they can indicate whether they want to repeat an experience or not. With food pellets as the reward, this is easily done as the pellets can be rationed on each trial so that the animal has to repeat the response if it wants more food. But to arrange repeated experiences with something like lying down or burrowing in a substrate or dustbathing is more difficult. The animal has somehow to be removed from what it has chosen the first time so that it can be offered another choice and this may mean being handled, pushed or shuttled away from what it has just chosen to see whether it still wants it the next time. This removal process, which is necessary for offering repeated choices, may itself be aversive to the animal, so that the methods used to find out what animals want can themselves affect both what they appear to want and how much they want it. For example, to show that young cows want to lie down enough to learn to operate a switch for the reward of lying down for 15 minutes, Jensen et al. (2004) had to devise a way of getting them to stand up again after 15 minutes so they could make another choice. So they put heifers into a harness that allowed the animals to move freely when they were standing up but could also be locked in place to prevent them from lying down until they had learnt to press a switch that released the harness lock and allowed the cow to lie down again. To train the cows, someone had to persuade the cow to stand up after each 15 minutes of lying time reward and lock the harness back to the hook ready for the next trial. Not everyone has the time, resources or ingenuity to carry out such operant tests and, in any case, being made to stand up every 15 minutes could itself have made the opportunity to lie down less rewarding for the cow.

While there are some situations, like the cow lying experiment, when giving an animal repeated experiences is just practically difficult and cumbersome, there are some situations where it is actually impossible for an animal to have the repeated experiences that are needed for it to give an 'informed' opinion of what is happening to it. For example, with a one-off medical procedure such as removal of a tumour, a question might be whether an animal (or a human) would prefer to have post-operative care with drug A or with drug B. It would be quite impossible to carry out same the operation repeatedly with different drugs and then see which one is chosen as the sequence of operations progressed. Fortunately, it is now possible to devise choice tests that avoid repeated testing but still ensure the animal has the relevant experience.

Conditioned Place Preference (CPP) is a way of finding out what animals want by training them to associate one place with a single experience such as food or a loud noise and a second place with something completely different, usually where nothing happens (Dixon et al. 2013). The two places are made very obviously different, such as being painted with different colours or patterns, to make it as easy as possible for the animal to associate each place with what happened to it there. The animal's preference for being in one place or another is measured both before and after its experiences in the two places.

If there is a shift in where the animal chooses to spend its time so that it starts to spend more time where it obtained the reward, this suggests that it liked the experience and is trying to repeat it (Bardo and Bevins 2000). Conversely, if it now avoids the place where the stimulus appeared and starts to prefer the place where it did not experience it, then this suggests that it found the stimulus aversive and wanted to get away from it. The advantage of this method is that it can work with just one experience, provided that the animal is clever enough to make the association.

CPP is commonly used in pharmacology to test how animals respond to different drugs (Bardo and Bevins 2000). For example, mice with bladder cancer show a preference for the place where they have been given morphine rather than where they have received saline, whereas healthy controls developed no such preference (Roughan et al. 2014). This suggests that the mice with cancer wanted the morphine.

Its opposite, Conditioned Place Aversion, has been demonstrated in a number of different species, including fish. Gilt-head bream, *Sparus aurata*, showed a conditioned avoidance to a place where they had been chased with a dip net compared with a distinctively different place in another part of their tank where they had not been chased (Millot et al. 2014). The fish showed that they did not want to be chased with a net by choosing the place where they had not been chased.

Indirect measures of choice—'out of sight'

Yet another way in which choice tests have been extended is to deal with what is sometimes referred to as the 'out of sight, out of mind' problem (Petherick et al. 1990; Warburton and Mason 2003). This is the possibility that an animal may start wanting something only when it is offered it in a choice test, so the choice test itself is altering what the animal wants. Some evidence for this comes from an experiment by Widowski and Duncan (2000), who trained hens to push open swing doors for the opportunity to dustbathe in peat. Some of the hens had been deprived of the opportunity to dustbathe before the tests and others had recently dustbathed; the idea being that the ones that had been deprived would want to dustbathe more and so would push open the doors more quickly. However, it turned out that there was no difference in how quickly the two groups of birds learnt to push open the closed doors. Both groups learnt readily and there was no observable difference in their behaviour when they could not see the peat, but once they had pushed open the doors and could see the peat, the deprived hens started dustbathing sooner. Widowski and Duncan suggest that this may mean the hens were stimulated to dustbathe only when they saw a suitable substrate (they all liked it when they saw it) but that they may not necessarily have wanted it when they could not see it (out of sight is out of mind). Warburton and Mason (2003) also found that, at least on some measures, how much mink apparently wanted access to resources such as waterbaths, food and social contact was affected by whether or not the animals could see or smell the options before making their choices.

Choice tests, in other words, present animals with opportunities that they do not have in the absence of those tests. Before seeing what is on offer, the animal might not want either option and only start wanting one of them when reminded by the choice test itself (in sight and in mind), rather in the way that you might only realize you were hungry if you smelt foods or saw someone else eating. Different rewards can also affect choice in different ways. A hungry animal choosing a food option, for example, becomes more likely to eat yet more food after taking the first mouthful (Day et al. 1997). Initially, eating food makes it hungrier and therefore even more likely to choose food the next time. Such positive feedback may not exist equally for all behaviour so this could distort the relative value that animals apparently put on different resources when compared with food. The problem that measuring what animals want by using a test that itself could alter what it is that they want has led to a search for new kinds of choice tests altogether.

Cognitive or judgement bias

A great deal of interest has recently grown in a kind of choice test that, on the face of it, appears to have little direct connection with what animals want or do not want. This indirect choice approach is based on the finding that humans who are depressed tend to make more 'pessimistic' judgements about ambiguous stimuli than people who are not depressed. For example, people who are depressed are more likely to interpret pairs of words such as die/dye or weak/week in their negative sense (Mogg et al. 2006) and people who are anxious about social situations are more likely to interpret an ambiguous facial expression as angry or otherwise negative (Richards et al. 2002).

Mendl et al. (2009) argued that such cognitive or judgement biases might also have important applications for understanding how non-human animals interpret the world and cited a study on rats to show one possible way in which this might be done (Harding et al. 2004). Rats were given a food reward if they pressed a lever on hearing a sound of a particular tone but not when they heard a sound with a different tone. When the rats had learnt this task, they were divided into two groups. One group lived in normal housing conditions while the other group were kept in barren cages. After 9 days of this, all the rats were retested with the two tones but this time they were also presented with sounds that were intermediate in tone between the two positive and negative ones. The two groups reacted differently to these intermediate or ambiguous tones. The rats that had been living in the barren environments were more likely to classify them as negative and not to press the lever, while the rats that had been living in the normal housing pressed the lever as if they had heard the positive tone. The rats from the barren environment thus showed a more negative judgement bias—that is, they treated the intermediate tones as not likely to be associated with food, like depressed humans taking a pessimistic view of a situation. By contrast, the normally housed group responded to those same intermediate tones as though they were a signal of food, more like optimistic humans 'looking on the bright side'.

Figure 4.6 shows in more detail how a 'cognitive bias' test is designed, this time using the place where an animal finds food as the positive stimulus and the place where there is no food as the negative stimulus (Burman et al., 2008). Twenty-four rats were all initially housed in enriched cages with sawdust, shredded paper as bedding and a variety of

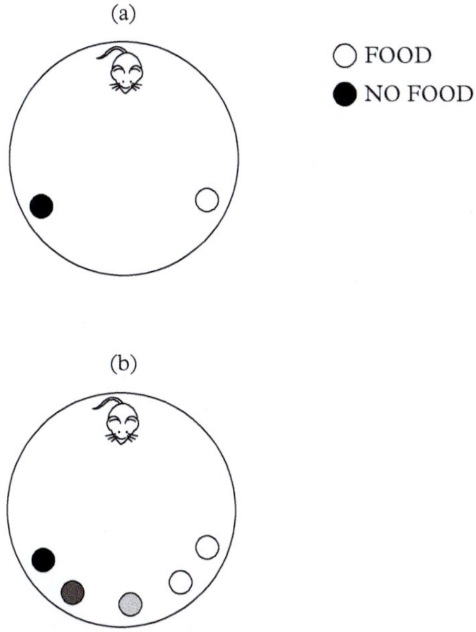

Figure 4.6 *Cognitive (judgement) bias experiments look for differences in responses to intermediate or ambiguous stimuli between animals that have had different experiences. Here is an experiment in which 'food' could be distinguished from 'no food' by where a food pot was located. Rats were first trained to go to one of two locations to find food (a). The food was hidden inside a specially designed pot that prevented them from seeing the food directly. To avoid the rats just using smell to find food, both pots contained food, but in the 'no food' location the food was made inaccessible by having a wire grid placed over it, whereas in the 'food' location, there was no grid and the rats could reach the food. Both pots therefore smelt of food but the rats had to look inside them to discover whether the grid was there or not. Once they had learnt that food was obtainable in one location but not in the other, they were given tests in which there were additional pots at intermediate locations (b). The question asked was whether the rats treated these intermediate pots as being near enough to the 'food' location to be seen as likely to contain food (i.e. they looked inside them as they would at the original food location) or whether they treated them as being nearer to the 'no food' location (i.e. they did not look inside expecting the food to be inaccessible). Redrawn from Burman et al. (2008).*

other objects such as a shelter. After 7 weeks, half of the rats had the enrichments removed from their cages, leaving them with only sawdust, food and water. The remaining 12 rats continued to live in the enriched cages. All the rats were then trained to find food in a pot at one location and no food in an identical pot at another location as shown in Figure 4.6. To control for the fact that the rats might simply be able to smell the food, both pots contained food, but in the 'no food' pot, the food was made inaccessible by putting it below a wire grid. The rats had to put their heads inside the pot to see if the food was covered by wire. All the rats learnt to reliably go to the place where food was accessible. They were then presented with more identical pots placed in positions intermediate between where the food and no food pots had previously been positioned. The aim was to see whether the rats regarded these intermediate pots as more like the positive stimulus (i.e. they would go to them and look inside as they would for a pot they expected to contain food) or more like the negative stimulus (i.e. they would ignore them as they did not expect them to contain food).

The results showed that, at least for the intermediate pot closest to the positive stimulus, the enriched group were more likely to treat it as if it were a positive stimulus, while the deprived group treated it more like a negative stimulus. Burman et al. (2008) interpreted this difference or 'bias' as evidence that the deprived rats were more pessimistic and that their mood had been depressed by the experience of having their enrichments taken away from them.

Depression in humans is a negative state (on the assumption that people do not want to be depressed) so, by analogy, animals showing a negative judgement bias are also seen as depressed and are, or recently have been, in a situation they do not want to be in. Negative judgement bias is seen as an indirect, but powerful, way of similarly finding out whether animals are, or have been, in situations that they do not want. Many different ways of conducting them have now been developed, many different species have been tested and the method has become widely used (e.g. Bateson and Matheson 2007; Bethell 2015; Baciadonna and McElliot 2015; Bateson 2016).

One major advantage of using cognitive or judgement bias rather than direct choice or operant conditioning is that it assesses the long-term effects of an animal either having or not having what it wants—that is, it assesses mood rather than immediate response to stimuli. It thus gets over the 'out of sight, out of mind' problem because the animal is not confronted with a direct choice that could change what it wants. Judgement biases focus instead on how animals interpret their environments when, for long periods of time, they are deprived of what they want or forced to live in conditions they do not want. They are thus potentially valuable but indirect measures of what animals want based on changes in the long-term state of the animals.

This indirect connection to what animals want, however, also has its disadvantages. Consider an animal that lives in an environment where food is only ever found in one place. The animal learns that it is pointless looking for food anywhere else because it is never there. Then imagine a second animal living in an environment in which food could be anywhere—under leaves, inside logs or in holes in the ground. When this second animal looks for food, sometimes it finds it and sometimes it does not, so it learns that it is always worth having a look. Both animals are then given a judgement bias test such as the one depicted in Figure 4.6 in which they have to learn that in Place A, there is a lot of food, whereas in place B there is no food. They are both tested for their response to intermediate places. It would not be surprising to find that the animal that had previously been in the environment where food was only ever found in one specific place would have a negative bias—that is, it would be less likely to explore the intermediate places than the 'scattered food' animal that had learnt that food might be found in a variety of different places. But the conclusion that living in a 'specific place' environment had been bad for welfare and caused the animal to become depressed and pessimistic would not be justified. It could simply be that the animal that had learnt that food is found only in one place had also learnt there is no point in looking anywhere else. The animal that had found food in many different places, on the other hand, had also learnt that intermediate places sometimes yielded food and so were always worth exploring. Being kept in barren (food delivered in a hopper) or enriched (food scattered everywhere) environment might well have given animals exactly these different sorts of experiences about where food is to be found and so any negative biases some of them might show say nothing in themselves about whether the animals are depressed (Houston et al. 2012; Bateson 2016).

In other words, because judgement bias is only an indirect measure of choice, its connection to what animals do or do not want, and therefore to welfare, is far from clear. Animals that have been living in conditions where they have not had what they want may indeed show relative negative judgement bias and such a bias could then be a reflection of the effect their environment has had on them. But the converse—that animals that show negative judgement bias must have been living in conditions which they did not like—does not necessarily follow. This would be like arguing that as all dogs are mammals, all mammals must be dogs. So it is now increasingly acknowledged that although judgement bias tests are a useful tool, they do need to be carefully interpreted (Bateson 2016) and backed up by other evidence (Bethell 2015). A negative judgement bias may suggest that an animal has been living in conditions it found aversive, but unless it is tied to more direct measures of what animals want, such as what it will work for, their interpretation will remain controversial.

Conclusions

We have now seen that there are many ways of asking animals what they want and how much they want it. They can vote with their feet, their paws and their beaks, and if provided with the right equipment, they can push switches, pull levers or use touch screens to indicate what they want. We have also seen that interpreting the choices that they make correctly takes care and patience. In particular, an animal's previous experiences—what it is familiar with—can affect what it appears to want. To this we should add that a wide variety of factors, such as which time in their lives they are tested, what motivational state they are in (e.g. whether they have just eaten) and the time of day, can all affect the choices they make. For example, pigs choose bright light when they want to defecate but prefer dimmer light when they sleep (Taylor et al. 1996). What animals want may also vary between individuals (Nicol et al. 2009, 2011), so there is no such thing as a single definitive choice (Fraser and Nicol 2011; Franks 2019) that holds for all members of a species.

While these points have to be kept in mind when designing choice or operant conditioning experiments, we should also remember that these problems of variation are not unique to animal choice, or even to animal welfare. Human medicine is only gradually coming to terms with the fact that different human bodies react differently to the same medication, but that does not invalidate the aim of finding cures that help most people. As we will see in Chapter 7, both humans and other species can show just as many differences between individuals in their physiological responses as they do in what they choose. An animal's physiological state also varies with time of day, whether it has just eaten or been chased by a predator as well as what other competing goals it might have.

Welfare measures, including what animals want, are part of biology, and all biological measures tend to be affected by many different factors. We just have to accept the inherent variability of biological systems and take into account the many different factors that could be affecting their behaviour, physiology and everything else about them before coming to conclusions about their welfare. We can still use what animals choose and what they will work for as key parts of assessing their welfare, but we do need to find out how these are affected by different factors, in the same way that 'personalized medicine' is now needed for the proper treatment of humans.

The practical problems with choice tests—in particular the difficulty of actually offering choices to animals—are, however, very real. Many, if not most, of the studies described in this chapter were carried out by academic researchers, often using special equipment to allow animals to express their 'opinions' that would simply not be available to most people with daily care of animals. The studies we have looked at so far are therefore best seen as essential background research rather than as the methods intended to be used by farmers, vets and pet owners on an everyday basis. What such people need—what anyone concerned with animal welfare needs—are faster, cheaper easier-to-use ways of judging whether animals do or do not have what they want. We will see what these are in the next chapter.

5

Behavioural Correlates of Welfare

So far, we have considered animal welfare in terms of just two measures—health and what animals want—and this may have struck some readers as rather a short list. What about all the other ways of assessing welfare that are now available, such as the sounds that animals make, their hormone levels, tendency to perform stereotyped behaviour and many more? The view taken in this book is that all these measures are important but that they should only be used as part of the total picture of 'good welfare' if they pass the test of being well correlated with either health or what animals want, or both.

In other words, health and what animals want are on a short list of their own for defining what animal welfare is. They take definitional priority over everything else and, furthermore, they have the power to validate any other proposed way of assessing welfare. On this view, all other indicators of welfare must prove their worth as the 'correlates' of health or what animals want.

What are 'correlates of welfare'?

'Correlates of welfare' do not define welfare. They are behaviours, sounds or physiological changes that are so closely correlated with either health or what animals want that they can be used as substitutes or diagnostics for the two core elements. For example, young piglets often utter high-pitched squeals. Underweight piglets and those that have not been fed for some time give longer and higher frequency calls (Weary and Fraser 1995). The piglet squealing calls are thus correlates of both health (as judged by piglet weight) and how much the piglets want food (as judged by how long since they have been fed). The high-pitched calls do not define what good welfare is, but they provide an easy way for judging the welfare state of the animals because they are well correlated with what does define welfare. There has to be considerable background research to establish the validity of a welfare correlate, in this case, showing that the calls do indeed correlate with both health and what the animals want. But once that basic research has been done, everyone else can simply go ahead and use the calls as 'correlates of welfare'. The squealing calls can be heard and interpreted by anyone without the need for special equipment. 'Correlates of welfare' thus have a very important role in the day-to-day assessment of welfare outside the research laboratory. They are what appear on

The Science of Animal Welfare: Understanding What Animals Want. Marian Stamp Dawkins, Oxford University Press (2021).
 DOI: 10.1093/oso/9780198848981.003.0005

welfare auditing lists, what farmers and stockpeople look out for, how animal welfare is generally understood in the outside world. But they get their status—their valence—from their correlation with either health or what animals want.

In the next three chapters, we will look at a wide variety of behavioural and physiological measures that have been used in the assessment of animal welfare. The view taken here—and not everyone will agree with this—is that many of the so called 'indicators' of welfare that are now widely used, such as stereotypies, stress, calls and grimaces, are best seen not as independent welfare measures in their own right but as correlates of welfare where their value has to be judged by how closely they are indicative of the two elements of the core definition. To illustrate this, we will look at some widely used indicators of welfare and challenge the validity of each one with two questions: Does it correlate with the state of an animal's health? And/or, does it correlate with whether animals have what they want?

Correlating behaviour with health and what animals want

Correlating behaviour with an animal's health is complex but relatively straightforward—we simply ask whether something such as a raised temperature or lethargy are reliable symptoms of a current or future disease state, with the full resources of veterinary medicine behind us. But correlating behaviour with what animals want is more difficult because animals do not have just one way of expressing that they have what they want and another, opposite, way of expressing that they do not have what they want. Different species have different ways of expressing what they want and even within one species, wanting one thing such as food will lead to quite different behavioural outcomes from wanting something else such as a safe place to hide from a predator. Similarly, animals that find themselves in situations that they want to escape from, such as being alone, will behave quite differently from ones that have, say, been used to finding food in a particular place and then suddenly find their food supply cut off. Attempts to escape and responses to not obtaining what is expected evoke quite different behavioural responses, even though both occur in situations with negative valence—that is, ones that the animals 'do not want' to be in.

A simple way of representing the complexity of the results of animals either having or not having what they want is shown in Figure 5.1. The vertical line shows a continuum from obtaining a reward to receiving a punishment so the highest welfare outcomes are at the top of the graph (positive valence). An animal positioned here has what it wants, whereas one at the bottom clearly does not. How much an animal wants or does not want something is represented by the distance from the centre of the line, which is therefore a neutral point of neither wanting to approach nor wanting to avoid (this single line thus represents both the valence and the intensity represented by the two axes in Figure 4.6).

But whether an animal can be described as having what it wants is also determined by the availabilities of rewards and punishments represented by the horizontal axis. Both rewards and punishments can be either present, absent, present but then withdrawn/unavailable or absent but impending. These various contingencies have very different

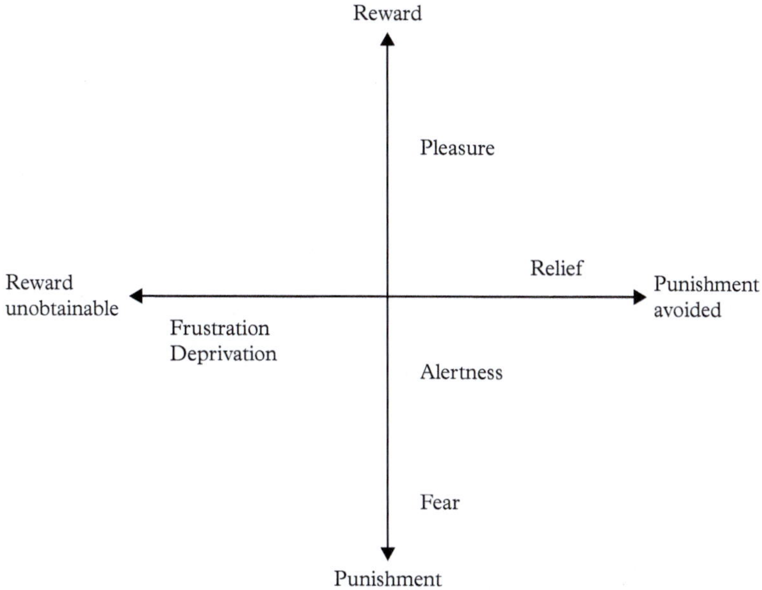

Figure 5.1 *Different ways in which animals can be in states of wanting or not wanting. The vertical axis describes a continuum from reward (associated with positive emotions when an animal has what it wants) at the top to punishment (associated with negative emotions when an animal 'suffers' from having what it does not want) at the bottom. The horizontal axis describes states associated with the non-delivery of expected reward (left) or the non-delivery of a punishment (right). The mid-point of each axis represents an emotionally neutral state, while distance from the mid-point of the line represents the intensity of wanting. The diagram does not imply a dimensional theory of emotion. Rather, it summarizes the states that could arise from a pair of reinforcers (one positive and one negative) and emphasizes the different behaviours that are expected under different contingencies. Different reinforcers will give rise to different states and therefore to different behaviours. Adapted from Rolls 2014.*

consequences for animals. For example, rewards can be present, but if they become unobtainable (if food suddenly stops or if a cow can see but not get to her calf), the animal does not have what it wants and is in a state of negative valence. At the far left are a cluster of such negative states, including being physically prevented from obtaining a reward that is in view ('thwarting'), failing to find a reward that was previously present ('frustration') and being unable to locate a reward that is wanted ('deprivation').

At the far right of the graph, we see yet more ways in which animals can be in a state of not having what they want. If a punishment is removed, this may be what the animal wants but it implies that something unpleasant (that it did not want) has been happening to it up to that point and, even if an actual punishment is avoided, the process of escaping or avoiding can itself have negative valence ('fear', 'anxiety').

There are thus many different circumstances in which animals do not have what they want and many different ways in which they show it, because their attempts to get what they want are potential solutions to very different problems. An animal kept in a barren cage with few resources (deprived), for example, needs to start searching for what it wants, whereas one that can see what it wants but cannot actually get to it (thwarting) might start physically trying to break down barriers, and one that is trying to get away from something it wants to avoid (fear) might hide or flee. Even more confusingly, a single animal may use the same behaviour both to get what it wants (such as running after prey) and also to get away from what it does not want (running away from a predator).

This is why putting everything through the test of whether it is what the animal wants is so important. It allows us to make sense of both the variation in how animals achieve what they want and also in the overlap between behaviour associated with wanting and that associated with wanting to avoid. What does everything that an animal wants have in common? Answer: that it will choose it and work for it, despite variations in how it might get there. What does everything that an animal does not want have in common? Answer: that it will choose and work to avoid being in contact with it, despite having different ways of actually implementing its escape or avoidance. The distinction between what animals want and what they do not want—the 'valence' or value they put on something—is what provides the constant anchor or touch-point for everything else. It is the behaviour that they do as a result that varies.

Establishing 'valence'

Although many, if not most, behaviours that animals show when they do not have what they want will be situation-specific ways of trying achieve what they want, there is one situation that provides an interesting and revealing exception. This is the interaction between offspring and their parents. Whereas adults by and large have to achieve what they want for themselves, young animals often rely on their parents to give them what they want. They therefore express their state of wanting something not by taking action themselves (as often they are physically not capable of doing this) but by signalling to their parents to do something for them. As a by-product, this provides us humans with a ready-made valence indicator in the form of behaviour that is reliably correlated with what the animal does or does not want.

The calls of domestic chicks give a clear example. In the presence of a wide range of positive reinforcers such as food, warmth and companions, young chicks give what is called a 'pleasure twitter' or 'cri de plaisir' (Collias and Joos 1953; Andrew 1964). These are soft rhythmic calls of about 4 kHz, repeated about five times a second (Wood-Gush 1971). This call provides an easy way for us humans to be informed that the chicks have everything they want because we can effectively eavesdrop on a call evolved to keep the hen informed that her chicks are safe and close by (Figure 5.2a). However, chicks do not give this call when they are isolated, hungry, cold or in other situations that are negatively reinforcing (Wood-Gush 1971). Under these circumstances, their calls take the form of loud ear-splitting 'distress' calls (Collias and Joos 1953; Andrew 1964), repeated two to four times a second, with a sweep of frequencies from 5 to 2 kHz. This call makes it easy for the hen to locate a lost chick and subsequently to provide the warmth and protection the chicks wants (Figure 5.2b).

Twitter calls and distress calls therefore come out well as generalizable valence indicators, applicable to a wide range of situations and distinctively different depending on whether the chick has or does not have what it wants. But such clear-cut correlations with valence may only exist because chick calls have evolved in the rather special context of a vulnerable offspring communicating with its parent. Here, it is in the interests of the chick to have clear and unequivocal 'welfare indicators' because its calls are its main method of persuading its parent to give it what it wants, and parents have, within limits, been selected to respond to the welfare needs of their offspring (Trivers 1974). Outside the parent–offspring kinship bubble, however, animals generally have to obtain what they want for themselves rather than expecting to get it from another individual. They therefore show a wide variety of different responses, depending on what it is they want and how they can best go about obtaining it. As a result we find that in most other situations, positive and negative valence indicators are correspondingly much more variable, making it much more difficult to identify clear-cut indicators of what animals do or do not want.

Figure 5.2 *(a) The expression of pleasure. Domestic chicks that are at the correct temperature, have food and water and the company of each other or the hen utter soft continuous 'pleasure twitters' as an expression that they have all they want. Photo credit: Kharkhan Oleg (Shutterstock). (b) The expression of distress. If a chick becomes isolated or cold, it gives loud 'distress' calls that have effect of making the hen locate it (Wood-Gush 1971). Photo credit: slowmotiongli.*

(b)

A good example of behaviour being only moderately well correlated with valence is the calls of rats, which have in fact been widely used as welfare indicators. Rats communicate with each other with high-frequency calls, most of which are above 20 kHz and so outside the range of human hearing. Calls between 35 and 70 kHz are given when rats choose to approach something they want and in anticipation of a reward (Knutson et al. 1999), social contact with other rats (McGinnis and Vakulenko 2003) and during sexual behaviour. They also give these high-frequency calls when being 'tickled' by humans, something they seem to want, as they will learn to perform an operant task for the reward of being tickled (Burgdorf and Panksepp 2001). This led Panksepp (2007) to describe the calls as equivalent to the 'laughing' of human children when being tickled and so an expression of pleasure or positive welfare. By contrast, the lower frequency (but still ultrasonic) calls of rats between 20 and 32 kHz are associated with situations that rats were thought to find aversive, such as being trapped by a predator (Blanchard et al. 1990) or in anticipation of punishment (Knutson et al. 1999). Rat calls have therefore appeared to be ideal 'correlates' of welfare.

However, as Makowska and Weary (2013) point out, there are some more recent findings that cast doubt on a simple distinction between high-pitched calls being given in the presence of rewards and lower-pitched calls being given in the presence of punishments. In particular, it has now been found that the 50 kHz calls are also given when rats are being euthanized with carbon dioxide, a situation that they will work to avoid (Niel and Weary 2006). It is not yet clear whether there is genuine overlap in in the situations in which the calls are given or whether we humans have not yet picked up some subtle distinctions in the meaning of the calls (Opiol et al. 2015).

A similar issue of whether calls are reliable indicators of valence has also been encountered with the vocalizations of other species. Leliveld et al. (2016) put young pigs through a series of either positive or negative conditioning trials with a view to using the calls to identify vocal correlates of valence. Each individual pig was repeatedly taken from its home pen to a test room. The pigs in the Positive group were given a reward, such as a small amount of apple sauce, a toy to play with or some straw, so that over a series of 11 such trials, they came to associate the test room with positive reinforcement. The pigs in the Negative group were given a mild punishment, such as being startled by a plastic bag being waved near them or someone suddenly rattling a can, while those in the Control group were the same room for the same length of time but were not given either kind of experience. The vocalizations made in the first 2 minutes of a test trial—while the pigs were anticipating their reward or punishment—were recorded and the results showed that both the Positive and the Negative groups vocalized less than the Control group, possibly because the conditioning procedure itself made them more attentive to what might happen next. Both the Positive and Negative groups produced 'grunts', but the Positive group produced more short, high-frequency grunts and fewer long, low-frequency grunts, while the Negative group showed the opposite effect. Despite these differences, however, there was a great deal of overlap between the grunts produced by all three groups and, statistically speaking, less than 12 per cent of the observed variation could be attributed to differences in what sort of conditioning the pigs had been given. In other words, someone trying to 'read' whether a pig was anticipating a positive or a negative reinforcer would be able to extract some information from the calls, but would still be wrong on many occasions. As correlates of what pigs want, their vocalizations, as we currently understand them, do not seem very reliable (Friel et al. 2019). Of course, this may just reflect the fact that we don't yet understand the complexity of pig vocalizations and that with more research they might still turn out to be usable correlates of welfare (Manteuffel et al. 2004).

Apart from interactions between parents and offspring, then, the calls that animals give show some correlation with whether that animal is in the presence of a positive or negative reinforcer, but there is often too much overlap between the two situations for us to be able to use those calls as reliable, unambiguous welfare indicators, at least until we have a better understanding of what those calls mean to the animals themselves. As a consequence, attention has been turned to other behaviours that might be more reliable indicators of valence.

Sometimes these are obvious. An animal that drinks when given water, builds a nest when given straw or dustbathes when given sand is showing that it has what it wants and likes it. Conversely, a bird that flutters incessantly against the bars of its cage is showing in no uncertain way that it wants to escape. At other times, it is possible to use more subtle behavioural indications of what animals are trying to do. Charles Darwin, in his book *The Expression of the Emotions in Man and Animals* (Darwin 1872) called this 'The

Principle of Serviceable Associated Habits'. He wrote: 'Certain complex actions are of direct or indirect service under certain states of the mind, in order to relieve or gratify certain sensations, desires etc.' (p. 28). Good examples are the expressions of disgust shown by a wide variety of different species, including human babies, monkeys and rats, when they have tasted something they do not like. Their mouths gape (Berridge 2000), an action of 'direct service' in ejecting the offending substance from the mouth and so helping to get rid of what is not wanted.

When animals are unable to avoid unpleasant or punishing stimuli altogether and are forced to put up with a state of chronic pain, the behavioural correlates of this state have now been documented for several species using 'Grimace scores', which focus on facial expressions or whole-body analysis to give an indication of the extent of the pain. The Rat Grimace Score (RGS) for instance, documents the facial expressions of rats with different types and degrees of pain (Figure 5.3), which are validated by comparing a rat's behaviour and expressions when it is given or is not given morphine to relieve pain (Sotocinal et al. 2011). Rats not given morphine have a higher RGS score than rats given morphine.

A Mouse Grimace Score (Langford et al. 2010), a Horse Grimace Score (Schraven et al. 2018) and comparable scores for other species are now in wide use for the assessment of pain. The measurement of the behaviour involved can be quite complex, requiring analysis of photos or videos, but it is generally agreed that, although these scores are useful and promising ways of assessing whether an animal is in pain, they do need more work for them to be reliable. An inherent problem is that prey animals including rats, mice and horses may be selected *not* to give out signs that they are in pain unless they physically cannot help it. Predators are always on the lookout for prey that are behaving oddly or look as though they are weak (Krause and Ruxton 2002), so behaving as a normal member of a flock or herd without giving out special signs of being in pain may be the best defence. The need to appear normal may be one of the reasons why on the Horse Grimace Scale it is difficult to distinguish a horse that is in pain from a horse that is 'sleeping' (Schraven et al. 2018).

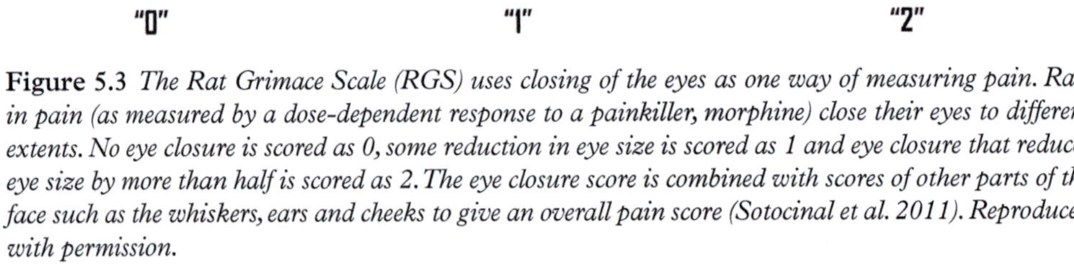

Figure 5.3 *The Rat Grimace Scale (RGS) uses closing of the eyes as one way of measuring pain. Rats in pain (as measured by a dose-dependent response to a painkiller, morphine) close their eyes to different extents. No eye closure is scored as 0, some reduction in eye size is scored as 1 and eye closure that reduces eye size by more than half is scored as 2. The eye closure score is combined with scores of other parts of the face such as the whiskers, ears and cheeks to give an overall pain score (Sotocinal et al. 2011). Reproduced with permission.*

In any case, good welfare is not just about the absence of pain or other punishments, it is about positive welfare (Boissy et al. 2007; Mellor 2016a, b; Webb et al. 2019), so what we need is the behavioural correlates of animals positively having what they want. Until recently, positive welfare has attracted far less attention than suffering and pain (Boissy et al. 2007), but this is possibly because it is much easier to recognize animals that do not have what they want than ones that do for the simple reason that an animal that has what it wants will not be making great efforts to change its circumstances or to escape from where it is. It is the animals that do not have what they want, or are cut off from companions, or are desperately trying to escape that will be behaving, calling and giving obvious signs of their state, and so providing us humans with behaviour that we can correlate with welfare. Contented animals have fewer incentives to take any action at all (McFarland 1985). They may be so calm, relaxed and content that they fail to give out any of the behavioural signs of positive welfare that we are looking for. It then becomes difficult to distinguish between an animal that is inactive because it is replete and content and one that has given up even trying to get what it wants. Because lack of activity is so difficult to interpret, and can even, as described above in the case of horses and birds, be indicative of pain, one approach to finding correlates of good welfare involves taking a step backwards in time from when an animal already has what it wants and focusing instead on the behaviour—often more active and much more obvious—of that same animal when it doesn't quite yet have what it wants but is eagerly anticipating that it soon will have.

Anticipation of reward or punishment

When animals are learning that a particular stimulus (or something they do themselves) is quickly followed by a reward or punishment, they soon start acting towards the stimulus as if the reinforcer itself were already available. Most people have heard of Pavlov's dogs, who learnt to associate the sound of a bell with food and started to salivate when they heard the bell on its own. Salivation is preparatory behaviour for eating (the chemical in it aid the digestion of food) so salivating in advance rather than waiting until food has actually arrived potentially speeds up digestion. Everywhere we look we see animals and people acting in preparation for situations that have yet to arrive and so being better prepared to make use of them when they do happen. From tennis players dancing from foot to foot in anticipation of the next serve to gulls following fishing boats, humans and animals act in preparation for what is to happen next or even to make something more likely to happen. Spruijt et al. (2001) suggested that what animals do during this period between a reward being signalled and the reward actually arriving is a period of 'positive anticipation' that can be a rich source of information about how animals express positive emotions. Conversely, the period between a punisher being signalled and its arrival is a period of 'negative anticipation' or fear. Either way, anticipation is a valuable way of animals showing us that they want (or do not want) something.

In anticipation of a reward, many animals including mink (Hansen and Jeppesen 2006) pigs (Imfeld-Mueller and Hillmann 2012), jungle fowl (Moe et al. 2014) and many zoo animals (Watters 2014) show a characteristic increase in activity and often vocalize more when they know that food is on its way, with predictable high-value food evoking more such behaviour than low-value or unpredictable food (Moe et al. 2014; McGrath et al. 2017). However, it is clearly critical to be able to distinguish anticipatory behaviour that animals show when expecting a reward from what may be very similar anticipatory behaviour shown when expecting a punishment. Very high levels of activity may also be shown by animals performing 'stereotyped' behaviour, which is, as we will see in the next section, widely regarded as a sign of stress or frustration. In all of these cases, the animal is getting ready for action and preparing to run towards something can look very much like preparing to run away from it. But movement alone has no valence. To give it valence, we need additional information in the form of whether the animal is moving because it is anticipating something it wants or because it is anticipating something it definitely does not want. To make sense of anticipatory movement in welfare terms, we need the animal's own view of what it is anticipating. This would be most convincing if we could identify behaviour that is consistently different depending on whether the animals are anticipating rewards or whether they are anticipating punishments (e.g. Moe et al. 2014; McGrath et al. 2017).

Behavioural diversity as a correlate of welfare

All animal behaviour can be seen as a constant series of decisions—to sleep, to eat, to drink, to mate—or on a finer time scale, to take another mouthful, to chew or to look around for predators. These decisions result in sequences of different actions that, for a human observer, can be described as more or less diverse (in the number of components) or as more or less predictable (in time). Both diversity and predictability have been widely used to give valence to behavioural correlates of welfare. But how well do these measures correlate with what animals want or do not want?

The diversity of an animals' behaviour—that is, the breadth of its behavioural repertoire—has been particularly important for welfare assessment in zoos and to a lesser extent in farms and laboratories (Haskell et al. 1996; Cronin and Ross 2019). Its use is based on the assumption that the more different behaviours an animal shows, the better its welfare must be. On the face of it, this assumption seems quite plausible. Captive animals are often kept in environments with restricted space and limited opportunities to perform some behaviour. That means that their behavioural repertoire can be expected to be more limited and less diverse than that of wild animals that, by contrast, have a richer environment in which to live and are able to exhibit a wider (more diverse) range of behaviour.

Two recent critiques (Cronin and Ross 2019; Howell and Cheyne 2019), however, point out the flaws in this assumption and argue that using behavioural diversity as a correlate of welfare is fraught with both theoretical and mathematical difficulties. For example, animals in captivity where there are no predators may show little in the way of anti-predator behaviour and in this sense they have an impoverished behavioural repertoire. But take those same animals and put them in a less safe environment where there are predators or threats of predators and suddenly a whole new range of behaviours becomes added. They become vigilant, they keep breaking off feeding to hide and so on. Their behavioural repertoire becomes much more diverse and yet it is not clear that their welfare has been in any way improved by exposure to danger. Conversely, an animal showing a restricted (less diverse) behavioural repertoire might simply be calm and contented because it has everything it wants around it and is not motivated to search or move much at all. Where a change in something can be equally well be interpreted as an indicator of good welfare or of poor welfare, it shows that it is not a valid behavioural correlate of welfare, at least if it is used in isolation. What is missing from measures of behavioural diversity taken on their own is that they do not indicate valence or the importance to the animal itself of being able to do a behaviour (Cronin and Ross 2019). In other words, behavioural diversity on its own does not show us what animals want or what they do not want. It needs to be reinforced by evidence of each individual element of the behavioural repertoire separately to find out which ones the animals themselves find rewarding enough to want to work to be able to do, which ones they are indifferent to and which ones they will work to avoid. Diversity alone does not stand up well to the two core questions about health and what the animal wants.

Behavioural predictability: stereotypies and other patterns in time

As well as being more or less diverse in its number of components, an animal's behaviour can also be described statistically by how rigidly fixed or flexible it is. In the assessment of animal welfare, a great deal of attention has been paid to one particular kind of fixed or rigid sequence that seems to be only seen in captive animals and may involve very high levels of activity. This is called 'stereotypic behaviour' (Meyer-Holtzapfel 1968; Mason 1991) and is characterized by an animal going through the same rigidly fixed, predictable sequence over and over again. Examples are 'weaving' in zoo elephants (swaying from side to side repeatedly (Greco et al. 2017)) and bar-biting in sows (moving the mouth backwards and forwards over the bars of a crate (Lawrence and Terlouw 1993). In captive polar bears, a common stereotypy is pacing, in which the animal traces and retraces the same fixed path for hours on end (Wechsler 1991) (Figure 5.4). Stereotypies are widely thought to denote poor welfare and are defined not just by their stereotyped or fixed nature (because that would include walking, breathing and other perfectly normal behaviours), but also because they are often performed in a frantic or abnormally rapid manner and appear to be quite functionless (Mason 1991).

Sometimes the link between stereotypies and welfare is clear and unequivocal. If animals actually injure themselves by repeatedly rubbing on the bars of a cage, there are damaging health implications and no-one is likely to argue that their welfare is not impaired. But it is becoming increasingly clear that stereotypies are not a single category of behaviour and that, on the contrary, they have a wide variety of different causes (Mason and Latham 2004; Nowak et al. 2016; Greco et al. 2017; Poirier and Bateson

Figure 5.4 *Polar bears (Ursus maritimus) in zoos often show stereotyped behaviour in the form of tracing and retracing the same path over and over again. The behaviour is so fixed (stereotyped) that the bear may take exactly the same a number of steps on each section of its path on each repetition (Wechsler 1991). Photo credit: Kari K (Shutterstock).*

2017). Stereotypies can be seen in animals anticipating the arrival of food (Hansen and Jeppesen 2006) as well as in confined caged animals. Sometimes animals even seem to benefit from doing them so they appear not to be functionless after all. For example, calves that have been separated from their mothers and fed from a bucket often show stereotyped sucking that may be directed towards other calves or inanimate objects. It appears to be functionless because the animals are repeatedly doing a behaviour that gives them no nourishment. De Passillé et al. (1993) provided such calves with a rubber teat that gave no milk at all but found that after the calves had drunk milk from a bucket, they would spend time in stereotypic and apparently pointless sucking on the teat. They then found, however, that the action of sucking stimulated the secretion of the hormones concerned with digestion such as insulin and cholecystokinin (CCK) and so improved the calves' digestion. Far from being a sign of poor welfare, the 'functionless' sucking was actually helping the calves to digest their food. As Mason and Latham (2004) put it, performing stereotypies may sometimes be an animal's way of providing itself with 'do-it-yourself enrichment'. It is also the case that some quite normal behaviour shows a high degree of stereotypy, such as the fixed way in which a chicken dips its beak in water and then raises its head to let the water trickle down its throat (Dawkins and Dawkins 1973).

So, as with behavioural diversity, stereotypy is not a valid welfare correlate on its own. Some stereotypies indicate poor welfare. Others show the opposite—they are part of the animal's coping mechanism and are no more indicative of poor welfare than a chicken drinking or a dog repeatedly chasing a ball on a beach. The problem with stereotypy is that, like behavioural diversity, it has no valence of its own. It does not tell us what animals want or don't want. Like diversity, stereotypy requires additional evidence before it can be used as a reliable indicator of welfare, specifically an investigation of the *value to the animal* of each individual 'stereotyped behaviour' on a case-by-case basis. Each stereotypy needs to be linked individually to either health or what the animal wants. Without such information, generalizations about the relationship between stereotypies and welfare are unconvincing and can be positively misleading.

Qualitative Behavioural Assessment (QBA)

In contrast to the quantitative approaches to animal welfare that involve collecting large amounts of data and then analysing it statistically that we have just been discussing, a much more intuitive approach to welfare assessment has recently gained popularity and has even found its way into the European Commissions' Welfare Quality® recommendations for welfare assessment. It is called Qualitative Behavioural Assessment (QBA) (Wemelsfelder et al. 2001) and it is exactly that—a qualitative, intuitive impression of how a human being interprets the behaviour of an animal. Observers are encouraged to look at the 'whole animal' and to use how it behaves, how it carries itself and its general demeanour as indications of its emotional state. The results are recorded as categories such as 'anxious', 'curious' and 'relaxed', these words being used in very much the same the way that they are applied to other human beings. Different humans tend to agree about which words should be applied to different animals (Phythian et al. 2013; Muri and Stubsjoen 2017) and in some ways this approach just formalizes what good vets, stockpeople and animal carers have always done in being alert to signs of ill-health in their animals, for example, by watching the way in which cows move and being able to detect those with mastitis (de Roches et al. 2018).

QBA is being increasingly used in as a method for assessing welfare, for example in sheltered dogs (Menchetti et al. 2019), in sheep during transport (Collins et al. 2018), broiler chickens (Muri et al. 2019) and in horses (Hintze et al. 2017), and it has the advantage over some of the other methods we have discussed that it claims to give 'valence' and to show whether an animal is in a positive or negative emotional state. However, it relies on human interpretation of animal body language and it does still need to be more directly related to what animals themselves want or do not want for it to be fully convincing. 'Valence' refers to what animals want, not to what humans think that animals want and human intuition needs to be constantly checked against the animals' own point of view.

Specific behavioural correlates of positive and negative anticipation

Given the difficulties of ascribing valence to broad categories of anticipatory behaviour such as activity levels, predictability of sequences or even stereotypies, it has often proved more useful to look for specific behaviours associated with anticipation of the different outcomes shown in Figure 5.1. For example, many animals have very characteristic behaviours associated with the anticipation of situations they want to avoid—what we call 'fear'. Signs of fear include increased vigilance in which the animal spends more time looking around (Welp et al. 2004), 'freezing' and remaining motionless, defaecation or actual flight. Many of these responses fall into Darwin's category of 'Serviceably Associated Habits' in that they help the animal to avoid the anticipated aversive event happening, such as by becoming more vigilant and so more likely to identify the source of danger. But because fear responses are adaptive, they also tend to be very different between species depending on what dangers that species is most likely to face and how they deal with them. The fear response to a predator of a bird that lives in open ground such as a partridge is to freeze, stay on the ground and rely on camouflage to avoid detection, while for a jungle fowl the fear response is more associated with actively running to cover. Yet other species of birds respond to danger by flying away. Understanding the behavioural correlates of negative anticipation in a given situation therefore requires a detailed understanding not just of each species but also what it is they are anticipating. Many species, including chickens, primates and others, take different actions depending on whether the negative situation they are anticipating is a flying predator coming from above or a ground predator coming at them from below (Manser 2013). The appearance of fear is very variable.

It is a similar picture with another major way in which animals can fail to get what they want; that is, when they positively anticipate a reward but are then prevented from obtaining it—what we call 'frustration' or 'thwarting' (Figure 5.1). For example, the extent to which cows show 'the whites of their eyes' has been suggested as a measurable correlate of how much they want something that they cannot have. When cows were shown something they wanted such as food or access to their calves but then prevented from obtaining it, they showed an increase in the percentage of visible white in their eyes (Sandem et al. 2002; Sandem and Braastad 2005). However, more recently, Lambert and Carter (2017) showed that this increase in percentage eye white is not specific to situations in which cows cannot have what they want. They observed the eyes of cows when they were given either high-energy feed (much liked by the cows), or woodchip (inedible and much disliked) or standard food as a neutral stimulus. They found that the percentage of eye white increased not only when the cows were given woodchip but also when they were given the high-energy food. The percentage of eye white visible was therefore not diagnostic of positive or negative valence but rather seemed to be more a symptom of excitement or arousal that could be either positive (pleasure at having the preferred food) or negative (having woodchip when high-energy food was expected). This certainly does not invalidate the idea of behavioural correlates of welfare, but it does show how careful we need to be in interpreting what animals do in terms of what they actually want. For each species, we need to test them in a variety of situations that they themselves have shown us they want or do not want before we can confidently call a behaviour a reliable correlate of welfare.

Play

A somewhat different approach to understanding what animals do when they have what they want is to use the behaviour that animals show when they have already satisfied their basic needs for food, water, physical comfort and security. Once they have everything they want, they may turn to doing less urgent but still important behaviours, such as grooming themselves, grooming each other or simply playing (Figure 5.5).

Particular interest has centred on the role of play as a behavioural correlate of good welfare on the grounds that if animals are taking time out to play, then that suggests that they already have what they want in other respects (Lawrence 1987; Boissy et al. 2007). This suggestion is supported by studies showing that play increases when other needs

Figure 5.5 *Play is commoner in young animals than adults. Eurasian brown bear cubs (Ursus arctos) playing while their mother looks on. Photo credit: Gedriius (Shutterstock).*

have been satisfied. For example, when wild meerkats (*Suricata suricatta*) were experimentally provided with extra food, they more than doubled their rate of play compared with non-provisioned controls (Sharpe et al. 2002). Conversely, when food becomes scarcer, a wide variety of animals including rhesus macaques (Loy 1970), rats (Siviy and Panksepp 1985) and dairy calves (Krachun et al. 2010) show a reduction in the amount of time they spend playing. If animals are injured or sick, they also play less but then start playing more when they recover (Fagen 1981). After being castrated, for instance, young lambs play noticeably less, but then after 3–4 days, start playing as much as before (Thornton and Waterman-Pearson 2002).

Play has therefore been seen as an indirect sign of good welfare based on the assumption that an animal that plays must be relaxed enough to be free from injury, hunger, sickness, predation risk and other immediate threats (Lawrence 1987; Boissy et al. 2007). Play is seen as a 'luxury' (Lawrence 1987) or 'low-resilience' behaviour (McFarland and Houston 1981), meaning that animals only do it when they already have most of what they want, rather in the way that people take up leisure activities such as sport when their basic survival needs have been met and they have the time and money to do other things.

What we need to know, however, is whether play is really indicative of animals having what they want and therefore qualifies as a usable 'behavioural correlate of welfare'. Held and Špinka (2011) pointed out that play sometimes occurs in situations where animals appear to be getting the opposite of what they want and even in situations of stress and conflict. For example, separating kittens from their mothers and reducing the amount of milk they received resulted in an increase in play (Bateson and Young 1981), as did a sudden decline in maternal care by rhesus monkeys on the birth of a younger sibling (Devinney et al. 2003). Galloping and bucking in calves when let out into a field after a period of confinement is a form of locomotor 'play' but appears to be a sign that the calves found their previous restriction aversive since control calves that had not been confined played less (Jensen 1999). Play occurring as a 'rebound' or relief may be a sign that the animal is at last able to behave in a way it has previously wanted to do but has been prevented from doing. Far from being a correlate of what the animal wants, play in this case may be a correlate of a previous prolonged period when the animal was prevented from doing what it wanted (Held and Špinka 2011).

On the other hand, one of the most fascinating things of all about play is that it can act as powerful reward in its own right. Animals appear to want to do it. Animals including rats (Siviy and Panksepp 1985) and chimpanzees (Mason et al. 1963) will work hard when the reward is nothing more than the opportunity to play. For dogs trained to search out drugs, the ones that are best at the task are not necessarily the ones with the most sensitive noses but those that regard the whole search as one big game in which 'work' is rewarded by a period of being able to play with a preferred person (Porritt et al. 2015; Hayes et al. 2018). But not all play is so rewarding (Held and Špinka 2011). Play fighting can escalate into dangerous real fighting that results in injury, and losing out in the rough and tumble of juvenile play may result in an animal spending its life low down in the adult social hierarchy (Bauer and Smuts 2007). We cannot therefore conclude that just because an animal is playing it is necessarily in a situation it wants to be or would choose to be in.

While play may sometimes be a correlate of good welfare, the difficulty of defining it in different species (Burghardt 2005), the diversity of situations in which it appears and its diverse consequences (Held and Špinka 2011; Ahloy-Dallaire et al. 2017) mean that it is currently difficult to interpret in welfare terms on its own. Before it can be confidently used as a behavioural correlate of welfare, play needs more research and more evidence that it is, in a given situation and for a given species, a genuine manifestation of a particular animal having what it wants.

Exploration, grooming and sleep

A variety of other behaviours, including the time taken to build a nest (Rock et al. 2014) exploration, grooming and sleep, have all been proposed as behavioural correlates of 'positive welfare' (Boissy et al. 2007). As this book is not intended as a comprehensive textbook, we will not go into them all here, but I am sure that by now you would have already anticipated the conclusion: for any new behavioural 'candidate' measure of welfare, it has to be able to unambiguously distinguish between animals having what they want and animals not having what they want. Any proposed 'welfare measure' such as exploration or sleep or response to novel objects or approach distance to humans or grooming has to be shown to be a reliable indicator of health or what the animal wants. In particular, it must unambiguously distinguish between positive and negative valence. Any measure that could equally well be interpreted as an indication that an animal has what it wants and also as an indication that it does not have what it wants is therefore not a valid measure of 'welfare' because it does not measure valence. It therefore has some use in telling us that an animal might be in a state of wanting something. It might even be a useful warning that a situation needs investigating for its welfare implications but it is not, on its own, a reliable correlate of welfare. There needs to be extra information about whether that animal is aroused because it is trying to escape or aroused because it is anticipating something it wants. To be valid as welfare indicators, behavioural correlates of welfare have to be just that—good correlates of health and what animals themselves want, and distinctly different between situations of positive and negative valence.

Behavioural correlates and the comparative method

A promising use of behavioural correlates as a way of asking captive animals what they want is to compare the behaviour of different species (Mason 2010). In a zoo, animals such as bears or wolves are confined in cages that give them far less space than they would have in the wild, raising questions about how much space such animals might want and even bigger questions about whether any zoo would be able to provide that amount of space. In practice, it is generally not feasible to offer a polar bear a choice between a large area and an even larger area and ask which one it prefers, so this is where imaginative use of behavioural correlates can do the next best thing. By examining which species had the highest levels of known correlates of poor welfare when their space was limited, it has been possible to work out which ones are least able to have what they want in captivity (Mellor et al. 2018). For example, Clubb and Mason (2003, 2004, 2007) collected data on the behavioural problems shown by 33 species of carnivores kept in zoos across the world. They found that the species that travelled the longest distances in the wild and had the largest body size were the most likely to develop the most pronounced locomotor stereotypies. Kroshko et al. (2016) subsequently showed that it was not so much the amount of locomotion an animal did in the wild that was important as its home range size. Clubb and Mason (2003) conclude that there might be some species that are simply not suitable for being kept in zoos because they are kept in conditions that are such a long way away from what they want, as revealed by the high levels of abnormal behaviour they show, whereas other species, being much less affected, might have much higher welfare in captivity.

Note the two-stage process in establishing this conclusion. First, the locomotor stereotypies used have to be validated as behavioural correlates of welfare by showing that they are genuinely an indication of frustration in this particular species—that is, they are genuinely indicative of the animal not having what it wants. 'Abnormal' here has to be shown to be associated with what the animal finds aversive. Then the validated behavioural correlate of welfare is in turn used to infer, indirectly, what it is that the animal wants (in this case more space) and how much it wants it. What the animal wants looms large in both steps. The method demands large amounts of data to be handled correctly to give a valid comparison between species, which may be one reason why it has been relatively little used so far (Mellor et al. 2018), but zoos do already collect very large amounts of data, so there is considerable potential for using this approach to test specific hypotheses about what improves or damages welfare. We can actually find out whether there are some species that can be given what they want in captivity and others that simply cannot.

Conclusions

The message of this chapter is that all proposed indicators of welfare need to be referred to the common currency or 'anchor' of what animals want and what keeps them healthy. They stand or fall by how well they are correlated with these two factors, and many of them fail because they are seen both when animals have what they want and also when they do not have what they want. They fail to provide valence, although they can indicate that an animal is aroused.

Controversies about whether a given behaviour is a 'good' measure of welfare do not have to end with unsatisfactory conclusions that sometimes it is and sometimes it isn't. The issue can be resolved by finding out whether the candidate 'measure' in question correlates well or poorly with what the animal wants. What animals choose, and particularly whether they find something rewarding or punishing, ultimately gives valence to all measures.

Progress in validating behavioural correlates of welfare in this way has been hampered by assumptions that at first sight appear perfectly reasonable, such as that animals that play must be in a state of good welfare, or that stereotypies are necessarily indicative of poor welfare. But if we rigorously apply the two criteria of health and what animals want, then it becomes clear that such assumptions are not necessarily valid. It also becomes clear what we have to do to put more valid assumptions in their place. In most cases, it has turned out that broad categories of behaviour such as 'activity', 'play', 'diversity' or 'stereotypy' are less useful as behavioural correlates of welfare than more specific behaviours that tell us how a particular species shows that it wants (or does not want) a particular result.

Even here, however, we need to be careful. Animals have many different ways of getting what they want and an even larger number of ways of showing that they do not like what they have, do not have what they want or can see what they want and cannot get to it. Learning to decipher their body language takes time and effort, but it is an effort well worth making. A future programme of research should give priority to documenting how different species express what they want and do not want, and how they do so in different situations. Documenting this will be key to improving the welfare of different species.

Vigilant readers may, however, have noticed one glaring omission from this list of proposed behavioural correlates of welfare. This is 'natural' behaviour, which has assumed great importance for the assessment of animal welfare not only among animal welfare scientists but also in the minds of almost everyone who cares about animal welfare. Natural behaviour features so prominently in discussions about animal welfare that it needs a chapter to itself.

6

Natural Behaviour

The word 'natural' has a particular resonance in the English language. Advertisers are fully aware of the power of this word and exploit it to the full in the marketing of everything from shampoo to butter. The fact that these products contain 'natural' ingredients is supposed to assure us of their goodness, their efficacy and their health-giving properties. Natural is good. How could it be otherwise? Anything 'unnatural' or 'artificial' is so obviously far less desirable and even dangerous that we are persuaded, without even needing to think about it, to choose what is natural.

'Natural' is particularly powerful when it comes to its partnership with the word 'behaviour'. Because 'natural' is good, it is assumed that natural behaviour must be good for animals too, so much so that the ability to perform natural behaviour has become for many people a necessary condition for good welfare (Yeates 2018). This in turn has led to priority being given to 'environmental enrichment', which often has the explicit aim of ensuring that animals in zoos, laboratories and farms are able to perform their 'natural behaviour' (Mellen and McPhee 2001; Rabin 2003). For some philosophers, natural behaviour assumes an even greater role and defines 'what the appropriate benchmark is for judging whether a given creature has decent opportunities for flourishing' (Nussbaum 2004, p. 310). So being able to perform natural behaviour has come to be seen as not just one of a long list of welfare correlates to be evaluated alongside a lot of others, it has become, for some, what good welfare *means.*

In this chapter, natural behaviour will be stripped of its pejorative, emotionally loaded aura and subjected to exactly the same two questions as all the other candidate welfare indicators we have considered so far. Does natural behaviour help to keep animals healthy? And is natural behaviour what animals want to do? Readers will then be able judge for themselves whether 'natural behaviour' really does deserve its place as a uniquely special way of approaching animal welfare.

Why natural behaviour?

'Natural behaviour' has a variety of meanings (Bracke and Hopster 2006; Špinka 2006) but in common usage it refers to the species-typical behaviour of animals shown when they are living in the places where their ancestors evolved or at least in man-made environments that allow them similar freedom of movement. What gives natural behaviour

The Science of Animal Welfare: Understanding What Animals Want. Marian Stamp Dawkins, Oxford University Press (2021).
 DOI: 10.1093/oso/9780198848981.003.0006

its special potency in many peoples' minds is that it is behaviour that an animal has been evolved to do. It is therefore seen as its rightful heritage—the essence of what that animal is when it is wild and free (note the use of two other emotionally loaded words here—'wild' and 'free'). Giving domesticated or captive animals the opportunity to behave naturally is therefore seen as giving them the opportunity to be themselves.

This emphasis on natural behaviour as key to animal welfare goes back a long way and one of its most persuasive proponents in the 1960s was W. H. Thorpe, who argued that ways of keeping animals that prevented them from expressing what he called their 'natural instinctive urges' undoubtedly caused suffering (1965). The Brambell Committee (Brambell 1965) agreed with Thorpe and concluded 'The degree to which the behavioural urges of the animal are frustrated under the particular conditions of the confinement must be a major consideration in determining its acceptability or otherwise' (p. 13).

The Farm Animal Welfare Council (FAWC 1979) subsequently took full account of the Brambell Committee's recommendations but carefully avoided the word 'natural' when formulating the Five Freedoms that it argued were needed for good welfare, preferring instead the word 'normal'. Thus the fourth of the Five Freedoms became 'The ability to perform most normal patterns of behaviour' (FAWC 1979, 2009). FAWC made this subtle but important distinction between 'natural' and 'normal' because of the difficulty of giving a definition of 'natural' for farm or other domesticated animals. 'Natural' is easy enough to define when it comes to a captured wild animal or even an animal born in a zoo. 'Natural' here means the behaviour shown by wild members of the same species. But for animals that have been domesticated for many generations and selectively bred to adapt to their human-made environments, it is far from obvious what their 'new natural' is. So FAWC's use of the word 'normal' was made in acknowledgement that, although many domesticated animals retain the ability to show much of the same behaviour as their wild ancestors, there may also be differences. Both by genetics and upbringing, what keeps domesticated animals healthy and what they want may have changed over the generations. The word 'normal' leaves room for this possibility.

The Fourth Freedom is the only one of the Five Freedoms to emphasize the positive aspects of welfare and to be explicit about what needs to be in place to give animals real freedom and quality of life. Freedom 4 looks beyond merely being free from hunger, thirst, injury and discomfort to what might be a really good life for animals, and it paved the way for more recent ideas about quality of life and a life worth living (Bono and Mori 2005; FAWC 2009; Broom 2007; Wathes 2010). As we saw in Chapter 2, good welfare has now come to mean much more than just the absence of suffering and ill-health. Not being hungry, thirsty or in pain is the bare minimum, but really good welfare should mean much more than this and be defined as an animal being in a positive affective state (FAWC 2009; Green and Mellor 2011; Webb et al. 2019; Yeates 2018). As we have seen throughout this book, positive affective state can be assessed by whether the animal has what it wants, so the crucial question for whether natural (or normal) behaviour should be used as an indication of positive welfare is whether natural (or normal) behaviour contributes to good health and whether it is what animals want to do.

The welfare of animals in nature

Life in the wild is not pleasant. Charles Darwin, in a letter to J. D. Hooker in 1856, wrote: 'What a book a devil's chaplain might write on the clumsy, wasteful, blundering, low and horribly cruel works of nature!' Wild animals are subject to constant threats of disease, not finding enough to eat or being eaten themselves. To judge by the standards of the Five Freedoms, they do not enjoy freedom from hunger, thirst, physical discomfort or physical injury. Captive animals are often healthier and live longer than their wild counterparts (Mason 2010). At least in captivity, an injured animal can be recognized and treated quickly, whereas a wild one could be left to a slow and lingering death. In the Serengeti National Park in Tanzania, large numbers of animals, particularly wildebeest, are often found dead and dying and the main cause of mortality (75 per cent of cases) is starvation and malnutrition, particularly if there has been low rainfall (Mduma et al. 1999). Through not having enough to eat, the bone marrow of the wildebeest deteriorates and they die in very poor condition. In other populations, such as white-tailed and mule deer, death by sickness and starvation is combined with heavy predation by bobcats, particularly on the fawns, so that in some years only 17 per cent of them survive their first year of life (Haskell et al. 2017).

Not only is life in the wild short and hazardous, but the method of death may be very painful. Leopards (*Panthera pardus*) for example, grip their prey by the throat so that they suffocate (Figure 6.1), and spotted hyenas (*Crocuta crocuta*) simply kill their victim by eating it. Kruuk (1972) describes in graphic details how a group of hyenas will attack the hindquarters of their prey, biting its loins and anal region, and some of them going for the throat and chest. They tear away chunks of skin, muscles and intestines from a living animal, and eventually bring it down and kill it in anything from 1 to 13 m.

The 'balance of nature', which sounds so harmonious and natural, is in fact achieved at a terrible cost to the welfare of individuals. If a population is not increasing over time and more young are produced than the single one needed to replace each adult, then inevitably that means that many animals are dying. This will be particularly true of species that produce large numbers of young because, in a balanced population, large numbers being born will also mean that equally large numbers are dying. Even in populations completely unaffected by humans, deaths by predation, disease, malnutrition, attacks by other animals, parasites and adverse weather happen all the time (White 2008; McCue 2010; Hill et al. 2019) and result in many wild animals having short lives in which, it has been argued, there is much more suffering than pleasure (Horta 2018).

Even when prey animals are not actually attacked or injured, their welfare may be compromised by the stress of having to be constantly on the lookout for danger and having to break off what they are doing and seek cover (Preisser et al. 2005). In a study of what they called the 'ecology of fear', MacLeod et al. (2018) showed that exposing female snowshoe hares (*Lepus americanus*) to the sight of a predator (a trained dog) shortened their lifespan and also reduced the survival of their offspring by over 85 per cent. Fear, anxiety and many of the negative emotional states associated with poor welfare are the features of everyday life for many wild animals in their natural state (Figure 6.2).

Figure 6.1 *Leopard (Panthera pardus) kill their prey by grabbing them by the throat until they suffocate. Photo credit: Mario bono (Shutterstock).*

Figure 6.2 *Fear is a natural occurrence in wild animals. Two spotted hyenas (Crocuta crocuta) have trapped an immature kudu (Tragelaphus). Photo credit: Albie Venter (Shutterstock).*

Many people will still argue, however, that, in spite of all the dangers it is still 'better' for animals to be free and subject to natural hazards than to be confined in a zoo, farm or laboratory. The implication here is that the wild animals are better off because they are doing what they want to do. That brings us to the really fundamental assumption that natural behaviour is indeed what animals want to do.

Do animals want to do natural (or normal) behaviour?

Not all natural behaviour makes a positive contribution to welfare. Being chased and caught by a predator, for example, is entirely natural for many wild animals but it is also something they will choose to avoid. It would therefore be difficult to argue that zoo animals should be regularly chased by predators because it is 'natural'. The only behaviours that contribute to welfare are those that either improve health or give animals what they want. There may be some behaviours that animals want to do and others that they do not.

Recognizing that not all natural behaviour contributes equally to welfare, Bracke and Hopster (2006) put forward a carefully modified definition of what 'natural behaviour' is and which aspects are most important to animal welfare. They argued that defining it simply as 'species-typical behaviour shown in nature' (i.e. by the wild cousins of our domestic and captive animals) is clearly not good enough. Natural behaviour only contributes to welfare if it is behaviour that animals will work to be able to perform it. In other words, their redefinition of natural behaviour as only including behaviour animals want to do enough to work for it is very much in line with the view of welfare taken in this book. It follows from their definition that there is some behaviour that that is quite natural in the sense that it is seen in wild animals but does not contribute to welfare because animals very definitely do not want to do it (Veasey et al. 1996; Fraser 2008; Wells 2009; Yeates 2018; Learmonth 2019). Being chased or harassed by predators is an obvious example. The naturalness of behaviour is therefore not what defines its contribution to welfare. Rather, it is the extent to which the animals themselves want to do it.

It is important to understand why animals may not always want to behave naturally. The fact that animals find some situations aversive or punishing is not some random by-product of evolution, put there to annoy us humans and frustrate our attempts to keep animals in farms or zoos. Animals want to avoid some natural situations because it was adaptive for their wild ancestors to be fearful or anxious when they were in danger. It was adaptive for those ancestors to seek out resources when they were deprived of them and to make efforts to regain the company of other members of their species if they had become isolated. It was adaptive to want to escape and to try new strategies if food supplies failed or water became scarce. The negative emotional states that we now associate with 'poor welfare'—fear, anger, frustration, deprivation, pain—were vital to the survival of their wild ancestors because they galvanized them into action and into changing their situations for the better. They helped to keep animals alive. Punishment and how to avoid it was as important to health, survival and future reproductive success as reward and how to obtain it. Finding themselves in situations they did not like and having the motivation and cognitive ability to get themselves out of such situations and into something better was vitally important. it was what made them able survive long enough to leave descendants, and those descendants—the animals whose welfare we are now concerned with—still carry the legacy of wanting to avoid certain perfectly natural situations because, to them, they are the harbingers of death and destruction. However natural, they are to be avoided at all costs. To have a truly animal-centred view of welfare, we need to respect that legacy and let the animals themselves tell us which natural behaviour they do want to do and which they do not, using techniques such as those discussed in the last two chapters.

Why natural behaviour is important

In the process of discovering what animals want to do, however, a knowledge of the natural or normal behavioural repertoire of a species is still a vital step in understanding what is good for their welfare. It draws our attention to the differences between wild and captive members of that species and therefore makes us aware of the possible behaviours that the captive ones *might* want to do. It does not say that they will necessarily want to do them just because they are natural but it provides obvious candidates to be tested. The fact that the jungle fowl ancestors of our domestic chickens always roost in trees at night, for example, highlights the possible importance of roosting to modern breeds. It does not tell us that all modern chickens still definitely want to roost, but it provides a very plausible hypothesis that this might be important to them. Such a hypothesis can then be tested by investigating whether modern chickens still want to roost (they do).

Understanding the natural environment in which animals have evolved, their evolutionary history and the mechanisms by which they achieve what they want can thus make an important contribution to establishing what good welfare involves for a particular species (Olsson and Keeling 2005; Mellor et al. 2018). Natural behaviour does not define good welfare on its own. It has to be validated in exactly the same way as all the other behavioural correlates of welfare.

Conclusions

Natural behaviour is important because it gives us a baseline and alerts us to possible welfare problem areas. It provides candidates for what animals might want that then can be investigated. It does not, however, define what welfare is. To link it firmly to 'welfare' it needs back-up information about what makes animals healthy and what they want. Animals do not necessarily want something just because it is natural.

Some natural behaviours will turn out to be positively reinforcing, while others will not. Natural behaviour has to be validated in exactly the same way as all the other behavioural correlates of welfare, as either contributing to health or being what the animals want to do. But it acts as a constant reminder that we are dealing with living beings that have evolved over millions of years into ways of life that are very different from ours and may have needs and wants that go way beyond our own particular human-bound way of looking at the world.

7

Physiological Correlates of Welfare

In the second half of the twentieth century, when the study of animal welfare was struggling to establish itself as a serious science, an obvious way to gain credibility was to draw on already well-established bodies of 'hard' science such as physiology and biochemistry (Fraser 2008; Veissier and Miele 2015). Good physiological functioning is obviously an integral part of animal health and, even more importantly, physiologists had recognizably scientific ways of documenting the responses of animals to 'stressors' such as heat, cold and injury through changes in their hormone levels, heart rate and other objective measures (Cannon 1929; Selye 1956). Linking physiological 'stress' to animal welfare therefore seemed to provide a way of delivering quantitative measures of welfare that would be widely recognized as scientific (Fraser 2008).

However, the interpretation of stress measurements has since proved to be one of the most controversial and indeed confusing aspects of animal welfare science (Dantzer and Mormède 1983; Rushen 1986b Mormède et al. 2007; Koolhaas et al. 2011). Since 'stress' carries with it connotations of being unpleasant and undesirable, then labelling something as a 'stress hormone' or a 'stress indicator' can lead unthinkingly to the conclusion that any increase at all in that hormone or other indicator shows that the animal must be more stressed and so its welfare must have declined. As we will see in this chapter, this is not always the case and highly pleasurable actions such as eating and sex can also be 'stressful' as judged by an increase in so-called stress indicators.

We will see that many physiological measures are extremely good at indicating that something is up—that the animal has been alerted or is attempting to take some sort of action—but not so good at indicating whether that alertness or motivation is part of wanting to escape from something it does not want or trying gain something it does want. In other words, physiological measurements are good for indicating arousal but not so good for indicating valence. As such, they are best seen not as measures of welfare in their own right but as correlates (and not always particularly reliable ones) of what we really want to know about welfare—namely, whether the animal's health is at risk or whether the animal does or does not have what it wants. Let us start by seeing why physiological 'stress' has been so important to animal welfare.

The Science of Animal Welfare: Understanding What Animals Want. Marian Stamp Dawkins, Oxford University Press (2021).
 DOI: 10.1093/oso/9780198848981.003.0007

Physiological symptoms of 'stress'

When an animal is suddenly faced with a situation in which it has to take some sort of rapid action such as running way from a predator or putting up a fight, its body changes to meet the emergency (Cannon 1929). It breathes deeply to take in extra oxygen; its heart beats faster so that the oxygen can be pumped more quickly to the tissues that need it. The liver releases sugar to fuel the muscles that will be used for fighting or running (Broom and Johnson 1993; Toates 1995). All this happens very quickly in a few seconds or minutes and is due to the complementary actions of the part of the nervous system known as the sympathetic nervous system and a cascade of hormones originating in the brain, spreading to the pituitary gland and then affecting more distant parts of the body, particularly the adrenal glands, which are situated just above the kidneys (Sapolsky 1994; Broom and Johnson 1993). The sympathetic nervous system stimulates the heart to beat faster, constricts the major arteries so that blood pressure goes up and increases blood flow to the muscles. The sympathetic nerve endings release the hormone norepinephrine (sometimes called noradrenaline) into the bloodstream and, at the same time, stimulate cells in the central part of the adrenal glands (known as the adrenal medulla) to produce another hormone called epinephrine (adrenalin) (Sapolsky 1994). The two hormones act together as chemical messengers, stimulating the body into a state where it is ready for action—heart beating fast, blood pumping, everything on full alert. This is sometime referred to as the SAM response (sympathetic and adrenal medulla response).

The SAM response occurs within seconds. If, however, the danger persists for longer than this, an additional set of hormones comes into play. The hypothalamus, situated at the base of the brain, secretes tiny amounts of a hormone called corticotrophic releasing hormone (CRH) directly into the anterior pituitary gland just below it. The pituitary gland in turn responds by releasing much larger quantities of another hormone called adreno-cortico-trophic hormone (ACTH, also known as corticotrophin) into the bloodstream. ACTH circulates in the blood and acts as a messenger to stimulate the adrenal glands into further action. This time it is the cortex or outer part of the adrenal glands that responds and the cells here start producing increasing amounts of steroid hormones called glucocorticoids, which come in a number of different forms. Birds, amphibians, reptiles and some mammals (mice, rats and rabbits) secrete mainly corticosterone, whereas in fish and almost all other mammals, the main hormone is cortisol (Koren et al. 2012). Collectively, these are often referred to as 'stress hormones' and although they act in slightly different ways, both corticosterone and cortisol have the effect of mobilizing stored energy so that it is available for powering action such as fast running (Vander et al. 1990; Sapolsky et al. 2000; Palme 2019).

The reason that glucocorticoids have been given the name 'stress hormones' is that they show a marked increase in the bloodstream following a wide variety of different aversive situations such as social separation (Boissy and LeNeindre 1997), injury (Lay et al. 1992; Sylvester et al. 1998), transport in calves (Kent and Ewbank 1983), being handled in trout (Pickering et al. 1982) and failure to obtain an expected reward in mink (Mason et al. 2001). They might seem, then, to be the universal indicators of situations that put animals under some sort of 'stress'.

Glucocorticoids have many effects on the body but one of the main things they do is to stimulate fat cells to give up their stored energy. They make the fat cells release fatty acids and glycerol into the bloodstream and they also trigger the conversion of glycogen to sugar in the form of glucose, which can be immediately used as muscle fuel. In addition, these hormones also cause any glucose that is already in the blood to be kept there by inhibiting the secretion of insulin from the pancreas. Insulin would normally be involved in removing glucose from the blood and storing it, but during the stress response, everything goes into reverse. Energy storage is temporarily halted so that muscles can be provided with immediately available fuel for fighting, running or other action (Sapolsky 1994). The pituitary gland joins in by secreting hormones that inhibit reproduction and growth, so that even more resources can be mobilized to meet the immediate emergency.

At this stage, the stress response can be described as adaptive because it maintains the body in a state of readiness to combat a source of danger. Usually, under natural conditions, when this alarm response is activated, a wild animal successfully escapes from its predator or fights off its opponent and can then resume less strenuous activities such as grooming or resting. The emergency is over and the nervous activity and hormone levels return to normal.

It may happen, however, that the emergency measures do not immediately result in the danger being removed and the animal remains in crisis mode. For example, if an animal cannot escape from the attacks of other animals, then it will be constantly in a state of preparation to escape but never able to actually get away. Its emergency reactions will be continuously aroused but to no avail. Eventually, if the danger continues, the animal enters the final or 'exhaustion' stage of the stress response and its adaptive mechanisms start to break down altogether. The thymus gland begins to shrink and fewer white blood cells (lymphocytes) are produced, which affects the animal's ability to resist infection, thus making it more susceptible to disease (Berghman 2016). Growth in young animals slows down and sexual activity declines. Gastric ulcers may appear, while the overworked adrenal glands, still producing hormones, become bigger and bigger (Broom and Johnson 1993; Sapolsky 1994).

This may sound paradoxical. If the stress response is adaptive and helps animals to get out of harm's way by enabling them to take vigorous action, why should it result in lost fertility, damage to the internal organs and deterioration of physical health? Why should physiological responses evolved to avoid death and ill-health end up themselves precipitating a greater likelihood of infection and other signs of sickness?

The answer is that the initial stages of the stress response are indeed adaptive and helpful to survival. Making the body ready to take immediate action can make the difference between life and death, but it comes at the cost of taking resources that would otherwise be used for growth, reproduction and even fighting off disease. Temporarily reducing reproduction, growth and immune responses thus frees up essential energy and nutrients for emergency use (Houston et al. 2007; McKean et al. 2008; Berghman 2016). Provided this is temporary, no lasting damage is done, but if the stress response is repeatedly turned on or if the animal cannot ever turn it off the stress response can eventually become as life-threatening as the external threat itself (Sapolsky 1994). The immune system is depressed so that the animal succumbs to infection and the repeated changes in hormone levels have long-term consequences for the cardiovascular, digestive, reproductive, immune and other systems. This of course is exactly the situation in which many captive animals find themselves. Confined in small cages and unable to escape the attacks of other members of their species or kept in barren environments where they do not have the resources they want such as shelter or water to swim in, their stress responses may be constantly activated and never turned off. Unlike most wild animals that become temporarily stressed, can take action and then become unstressed again, animals kept in unnatural environments often cannot take the action that they are motivated to take and so never 'de-stress'. It isn't that stress is necessarily bad for welfare. Some stress—and the behaviour it gives rise to—is essential for survival. But stress that is constant or never reduced leads to the diseases of chronic stress, such as gastric ulcers, damaged arteries and increased susceptibility to disease. And that has been the source of controversy. If some stress is adaptive but too much is damaging, how do we know where to draw the line between stress that indicates good welfare and stress that indicates poor welfare? Following the same logic as we have used throughout this book we can attempt to answer this question by first asking how well stress correlates with physical health and how well it correlates with what animals want.

Does stress correlate with physical health?

Selye (1956) referred to the whole sequence of responses, from the initial alarm reaction to the stage of resistance right through to the exhaustion stage as 'stress' or the general adaptation syndrome. The two extremes of the stress response are relatively easy to interpret in health terms. In the early stages, the animal is simply alert to possible dangers in its environment just as most wild animals are. Indeed 'alertness' is one of the signs of positive good health. The early stages of the stress response, therefore, suggest good welfare by a healthy animal taking appropriate action. By contrast, the later stages of the stress sequence clearly indicate poor physical health. Gastric ulcers, abnormally large adrenal glands and an inability to fight disease because of a compromised immune system are unequivocal symptoms of poor physical health (Barnett 1987). The animal is ill and its welfare severely compromised on these grounds alone. What has been of concern to animal welfare scientists is where, between these two extremes, adaptive responses turn into poor health.

Moberg (1985, 1987) argued that stress responses are so intimately bound up with health that it is possible to identify intermediate 'pre-pathological states' that indicate poor welfare before an animal is classified as clinically sick. Pre-pathological states, Moberg believed, could be recognized by a variety of abnormal symptoms such as depressed immune function and reduced reproduction that indicate that animals are *at risk* of future ill-health even if they currently appear quite healthy (Gross and Siegel 1981; Moberg 1985; Broom and Johnson 1993; Cockram and Hughes 2011). The issue then becomes whether these sub-clinical stress symptoms are sufficiently good predictors of later disease states that we can justifiably call them physiological correlates of poor health.

Barnett and Hemsworth (1990) argued that, for pigs, a sustained level of 40 per cent of glucocorticosteroid hormones in the blood above the normal baseline was an example of pre-pathological stress on the grounds that if this level was sustained for any length of time, it led to damaging physiological consequences. By 'damaging consequences' they meant a loss of health and fitness, and they argued that the 40 per cent increase in corticosteroid or 'stress hormone' levels showed that the animals had reached the point where, if such levels were continued, pigs would start to have lower survival prospects and decreased reproductive success in the long term.

However, this did not satisfy Mendl (1991), who argued that there was not enough good evidence linking short-term measures of hormone levels to long-term measures of fitness and that the whole idea of a 'welfare threshold' did not make sense because different elements of the stress response gave different answers as to where the threshold leading to poor physical health should be drawn. Nearly 30 years later, the connection between glucocorticoid levels and fitness is still unclear (Breuner and Berk 2019). Rushen (1991) pointed out that the term 'pre-pathological' was itself very imprecise because the symptoms could vary depending on the circumstances and what future pathology was being predicted.

In an attempt to get away from specific welfare thresholds altogether, Fraser and Broom (1990) proposed that poor welfare could be more usefully recognized by whether or not an animal was 'coping' with the external pressures put on it. By 'coping' they meant that, although an animal may be temporarily disturbed by, say, being made too cold or subjected to infection, it has the capacity to respond by warming itself up or mounting an immune reaction and can therefore quickly return to the state it was in before the disturbance happened. The capacity to respond to but then to be able to cope with a disturbance is sometimes referred to as 'resilience' (Russo et al. 2012; Colditz and Hine 2016) and carries with it the implication that as long as an animal is coping, it is within its normal range of adaptive stress responses and its health and welfare are therefore maintained. Only if the animal is pushed beyond its capacity to cope will its health break down and its welfare be compromised.

Although the idea of coping or resilience appears to be a way of making the link between 'stress' and 'welfare', in practice, no clear boundary has yet been identified between normal maintenance of physiological equilibrium and 'stress' responses that the animal cannot cope with (Rushen 1986b; Cockram 2004; Colditz and Hine 2016). The relationship between stress, resilience and physical health is thus still not straightforward. As an additional complication, the suite of symptoms that make up the stress response is now known to be much more varied than Selye originally thought, making the concept of a single all-purpose concept of 'stress' even less plausible.

Specifically, Selye (1974) believed that the body responds in the same general way regardless of whether the 'stressor' is cold or heat or attack by another animal. He described the way in which an animal's body responds to factors that are known to produce physical damage, such as starvation, infection or injury and believed the body reacts in the same ways to a wide range of situations, only some of which cause overt damage. This was an idea that has had great appeal to animal welfare scientists as it suggests that there might be universal measures of 'poor welfare' that could be used in a variety of situations even when animals appeared to be perfectly healthy. For example, understanding the hormonal stress responses of animals subjected to being too hot or too cold could then be used to see whether similar responses are shown in animals that are, say, being transported in a truck. If their hormonal profile were similar, the implication would be that the animals were similarly 'stressed' by the transport. Glucocorticoids—already as we have seen known 'stress' hormones—seemed prime candidates for this role of universal indicators of poor welfare.

The idea of a simple hormone test that could 'measure welfare' was (and is) deeply seductive to animal welfare scientists—who see the enticing possibility of having an objective biochemical way of quantifying degrees of distress in animals. To add to their attraction as welfare indicators, glucocorticoids can now be sampled non-invasively, without the need for catching the animal to take a blood sample. Glucocorticoids or their breakdown products are secreted into urine, faeces, milk and saliva (Palme 2019) which can be collected much more easily than blood and sampled repeatedly to give a stress profile over time. Hair and feathers also accumulate glucocorticoids and have been seen as providing a record of the cumulative corticosteroid secretion during the period when the hair or feathers were growing (Heimeberge et al. 2019; Will et al. 2019). These completely non-invasive ways of collecting glucocorticoid samples have opened up new possibilities for measuring hormone levels in farm, zoo and even completely wild animals (Sheriff et al. 2011; Palme 2019).

However, although there are common features in responses to different stressors, there are also important differences and many inconsistencies (Mason 1974; Rushen 1986b Koolhaas et al. 2011; Palme 2019). In particular, the milder forms of stress often have different hormonal 'signatures' depending on what exactly is causing the stress (Sapolsky 1994). For example, in rats, subordinate animals that are repeatedly subject to challenges from other rats but still fighting back and trying to establish their social position within the group, show increased sympathetic nervous activity, but subordinates that have given up and are just trying to escape have stress responses dominated by higher levels of glucocorticoid activity (Henry 1977).

An additional problem is that the level of glucocorticoids appearing in the blood is highly dependent on how long after the stressor appeared a sample is taken. Baseline activity of the HPA axis—that is, when the animal is calm and not subject to any threats or challenges—is highly variable at different times of day, with a natural peak in activity towards the end of the night in diurnal animals. Even when an external threat has appeared, the blood levels of glucocorticoids do not increase until about 10 minutes later, after which they may remain high for an hour or so and then decline again. Completely different conclusions could therefore be drawn depending on whether the blood samples are taken at night or during the day and whether they were taken immediately, after 20 minutes, after 2 hours or a week later (Koolhaas et al. 1997). For this reason, taking repeated samples over time and using a cumulative measure of the total response is sometimes used to give a more accurate picture of how the animal is responding (Koolhaas et al. 2011).

But there are inconsistencies, too, in whether longer-term stress is associated with a decrease or an increase in glucocorticoid levels. Although stress is commonly associated with increased levels of these hormones, there are other cases where it is associated with a decrease. Horses with sub-clinical spinal problems or anaemia were found to have lower levels of cortisol in both their blood and faeces than healthy horses (Pawluski et al. 2017). Wild-caught starlings (*Sturnus vulgaris*) that were repeatedly subjected to stressors such as loud noises, having their cages tapped or rolled and being placed in bird bags also showed a decrease in total corticosterone over a period of 18 days (Cyr et al. 2007).

There are also some puzzling anomalies between glucocorticoids and other measures of welfare, particularly behaviour. Calves that were restricted socially and spatially were shown to have levels of glucocorticoids similar to those of unrestricted calves, despite high behavioural motivation to have contact with other animals and to move around (Friend et al. 1985). Laying hens given access to an area containing enrichments such as grass and litter showed a greater increase in glucocorticoids measured from their faeces than hens given access to a comparable 'barren' area with a wire floor, even though the hens were more attracted to the enriched areas (Dawkins et al. 2004 and Figure 7.1). One possible explanation is that the hens given access to the enriched area were more active and this was reflected in higher levels of glucocorticoids. This is supported by a study of rats that showed a rise in plasma corticosterone, heart rate and blood pressure when the animals were exposed to a novel environment and started actively exploring it. The physiological changes appeared to be the perfectly normal way their bodies were responding to greater exploration and movement rather than an indication of stress (Beerling et al. 2011).

The possibility that a rise in corticosteroid hormones may simply be associated with increased activity rather than with any negative state we would want to call 'stress' brings us to the biggest problem of all with relating 'stress' hormones to animal welfare. There

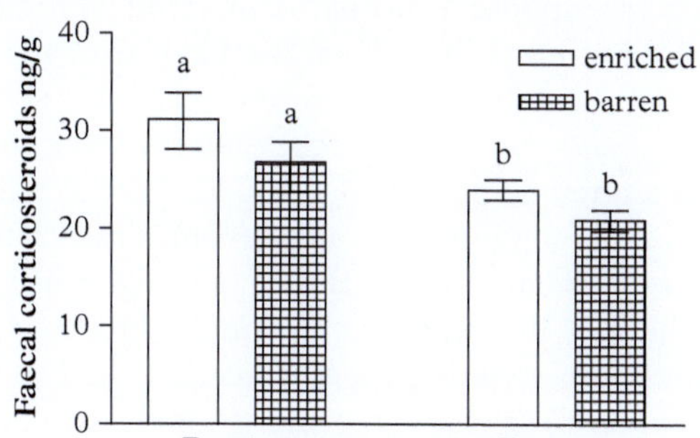

Figure 7.1 *A preferred environment can increase 'stress'. Hens show higher levels of 'stress' hormone when given access to their preferred environment. One group of hens was given access to a barren wire-floored cage with no enrichments and another group was given access to a cage with enrichments including grass and a scratching area. The birds were tested over a period of a week to establish where they chose to spend their time and faecal samples were collected on the first day of the test (Day 1) and again on Day 4. Redrawn from Dawkins et al. (2004).*

Figure 7.2 *Similar responses, different valence. Both a predator anticipating a meal and its fleeing prey show the same behaviour—running—and both show similar physiological adaptations for action. However, the cheetah (Acinonyx jubatus) runs in anticipation of a reward (food) while the warthog (Phacochoerus africanus) runs to avoid injury and death. Photo credit: Dennis W Donohue (Shutterstock).*

is increasing evidence that glucocorticoids increase not only in negative situations that animals find aversive but also in positive situations that they find highly pleasurable.

Does stress correlate with what animals want?

As we have seen, the 'stress response' is primarily involved with mobilizing the body for action and glucocorticoid hormones have been used as one of the major indicators that this process is under way. Two of the most common emergency actions that animals take are fleeing from a predator and putting up a fight against a rival, which are stressful both in the scientific sense of a physiological change and also in the colloquial sense of being undesirable or negative. 'Stress' therefore easily becomes associated in our minds with negative emotions such as 'fear' and 'anger'. But equally vigorous action is also called for when an animal is not the prey fleeing for its life but the predator running hard after its first meal for several days. The sprinting cheetah needs to fuel its muscles just as much as the fleeing warthog (Figure 7.2). Both are taking emergency action and both will be showing many of the same 'stress responses' as preparation for their exertions. The difference is that the stress for the cheetah is the pleasurable anticipation of a meal, while for the warthog the stress is the fear of being caught.

Many energetically demanding but rewarding behaviours, such as sexual activity (Bronson and Desjardins 1982; Bonilla-Jaime et al. 2006; Ralph and Tillbrook 2016), voluntary exercise (Stranahan et al. 2008) and winning a social encounter (Koolhaas et al. 2011) have been shown to involve large increases in glucocorticoid levels and other measures of so-called 'stress'. In rats exposed to a variety of 'stressors', by far the highest

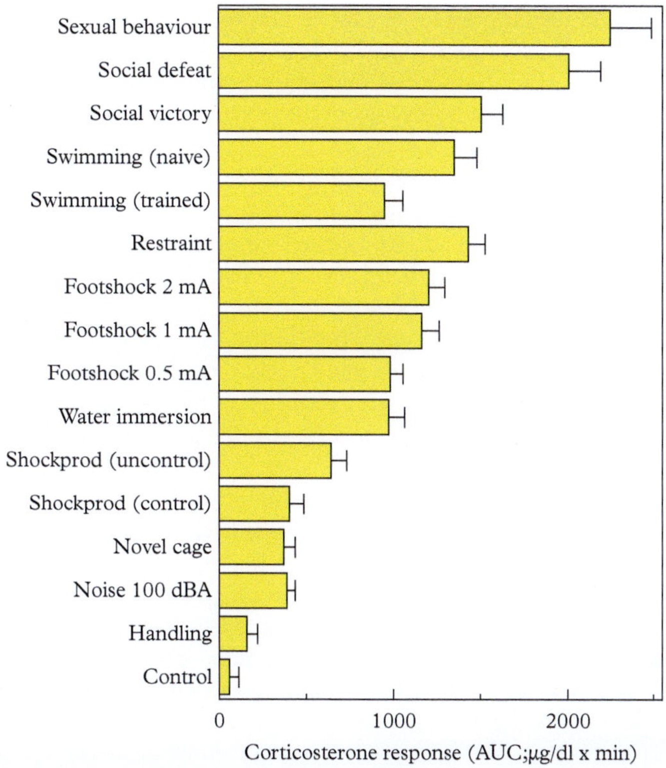

Figure 7.3 *Sex is 'stressful'. Adult male rats were exposed to a variety of different stimuli, such as being handled, having sex with a female or given an electric shock to the foot. Each test consisted of a baseline measurement, then 15 minutes of exposure to the stimulus, followed by 1 hour's recovery time, and their total amount of plasma corticosterone levels over this period was measured. The hormone levels were higher when the rats had been engaged in sexual activity than in any of the other tests. Redrawn from Koolhaas et al. 2011.*

levels of glucocorticoid hormone levels occurred during sexual activity (Figure 7.3), followed by both winning and losing a social encounter. Being given an electric shock on the foot or being handled evoked very much lower responses and there seemed to be no correlation at all between the levels of glucocorticoids and what the animals wanted to do (Koolhaas et al. 2011). Male rats confronted with an aggressive male intruder or a receptive female showed very similar corticosterone responses and their levels of plasma corticosterone were correlated not what the stimulus was but with the amount of behavioural activity they were showing (Buwalda et al. 2012).

What these results show is that it is impossible to tell, from hormone levels alone, whether an animal is finding a stimulus attractive or aversive (Ralph and Tillbrook 2016).

Simply eating food leads to a rise in corticosteroid levels as do many other normal body functions such as exploration (Olsson and Sapolsky 2006; Beerling et al. 2011). A rise in corticosteroid levels is part of necessary body maintenance or physiological preparations for activity—any activity—and therefore not uniquely indicative of either good or bad welfare.

In summary, the hope that measurements of stress in general or glucocorticoid hormones in particular could be used in any simple way as physiological correlates of welfare has to be treated with great caution. In extreme cases, that is, when an animal is in the exhaustion phase of stress, there is an obvious correlation with breakdown in physical health. Indeed, they are themselves the symptoms of disease. Malfunctioning organs and susceptibility to disease unequivocally point to severely reduced welfare. But in the earlier stages of 'alert' and 'resistance', the diversity of measures of stress used and the variety of different outcomes that can result mean that the correlations with one of the two key welfare measures—physical health—are far from clear. In addition, the fact that situations that animals find reinforcing, such as food and sex, also elicit similar responses to situations they find aversive shows that physiological measures of stress are not well correlated with the other key measure of welfare—what the animals want. Physiological measures of stress *on their own* are thus unreliable correlates of welfare. They are excellent measures of arousal but they do not provide valence—whether something is seen as positive or negative by the animal itself. Animal welfare scientists are therefore, often reluctantly, having to abandon the idea of being able to measure poor welfare in terms as simple as nanograms of stress hormone (Ralph and Tillbrook 2016; Palme 2019).

'Pleasure' hormones

The unreliability of stress hormones has led to a growing interest in the opposite endeavour—to find hormonal measurements of good welfare. With an increasing emphasis on good welfare as a state in which animals have positive emotions, not just absence of negative ones (Boissy et al. 2007), there has been a corresponding search for physiological correlates of pleasure and positive valence. Several suggestions have been made for this role.

Chief among these is oxytocin, a hormone and neuropeptide produced by the hypothalamus and released by the pituitary gland into the bloodstream. It can subsequently be detected in blood plasma, urine, saliva and milk, and has been called the 'love' or 'feel-good' hormone because of its involvement with pair bonding and maternal behaviour (Young and Zingg 2017). It appears to increase trust and reduce anxiety (Mitsui et al. 2011), and so has been suggested as an indicator of pleasure and positive welfare, particularly in social situations (Chen and Sato 2017). In support of this idea, oxytocin has been shown to increase in positive situations such as when a dog is being stroked by a familiar person (Rehn et al. 2014) or when cows are given food (Svennersten et al. 1990).

However, to be valid as a physiological correlate of welfare, it is not enough just to show that oxytocin increases in positive situations where animals have what they want. It has also to be demonstrated that it is not shown in neutral or negative situations where they want to avoid something or escape. Here the shortcomings of oxytocin as a welfare indicator become apparent. There is now evidence that punishments as well as rewards can lead to an increase in oxytocin (Rault et al. 2017). For example, rats subjected to 10 minutes of being shaken showed an increase in oxytocin (Nishioka et al. 1998) as did heifer calves showing a high degree of fearfulness (defaecation, avoidance of novel objects, lack of exploration (Yayou et al. 2010). Furthermore, injections of oxytocin actually lead to an increase in aggression in pigs (Rault et al. 2015).

As a physiological correlate of what animals want and find pleasurable, then, oxytocin does not stand up well. It is not as specific to situations that animals find positive or rewarding as sometimes thought and its normal functioning in the body is not well understood (Rault et al. 2017). One suggestion is that, rather being a sign of obvious 'pleasure' or 'liking', oxytocin should be seen as acting primarily to reduce sensitivity to non-rewards—that is, to reduce the negative impact of not getting what is wanted or expected (Ide et al. 2018). On this interpretation, its main role would be to enable animals to cope better with otherwise negative situations, raising further questions about its status as a welfare indicator.

Dopamine is another molecule that, like oxytocin, has complex roles within the body and functions both as a hormone and as a neurotransmitter. It, too, is associated with reward and pleasure. It is released from neurons whose cell bodies lie in the brainstem but which also connect to the basal ganglia and into the cortex. However, dopamine neurons do not encode pleasure or reward value per se. Rather they respond if the value of the reward is either greater or less than expected (Schultz 2013). They thus seem to be connected with learning rather than registering 'liking'. An even greater problem with using activity in these neurons as an indication of pleasure is that dopamine neurons come in many different types, only some of which respond to rewards. Others respond to aversive stimuli, such as a bitter taste (Schultz 2016), to novel stimuli (Bromberg-Martin et al. 2010) or even just the uncertainty of a reward (Fiorillo et al. 2003).

The idea of measurable, quantifiable, biochemical indicators of 'welfare' is, for many animal welfare scientists, the Holy Grail of welfare research—an ultimate goal that would solve many problems at a stroke. That is why the use of 'stress' hormones continues to have such appeal, despite its known problems that we discussed above (Ralph and Tillbrook 2016; Palme 2019). It is also why the idea of 'pleasure' hormones is even more seductive, enticing us to think that we can, just by measuring a hormone or a brain neurotransmitter, ensure that we give animals a life full of rewards and pleasure.

But, as Rault et al. (2017) point out in their review of oxytocin as a positive welfare indicator, entirely the wrong conclusions can be drawn unless it is first shown that results are consistent in giving one set of results to positive situations animals find rewarding and a different set of results to negative situations they find aversive. As of now, hormones given the labels of 'pleasure' or 'feel-good' such as oxytocin, serotonin and dopamine do not yet give a good enough indication of valence or what animals want that they can be relied on as physiological correlates of welfare. They may do one day as our knowledge of them increases, but as their use in welfare studies is still in its infancy, it is important not to oversimplify the many different roles they play within the body and the complex relationships they have with reward, punishment and arousal. They do not yet tell us reliably that an animal has what it wants.

Heart rate variability

Not all potential physiological correlates of welfare are hormones. As we saw earlier, another element of the 'stress' response is an increase in heart rate. This can be easily measured as the number of heart beats per minute, but as new technology has become available, more and more interest has centred on another aspect of heart rate: its variability (von Borell et al. 2007). The interval between one heart beat and the next is constantly varying over time but only by tiny amounts, measured in milliseconds. As you breathe out, the interval becomes a few milliseconds longer than when you breathe in. This constant adjustment of the heart rate is the result of a delicate balance between the two parts of the autonomic nervous system, the sympathetic and the parasympathetic (Bootsma et al. 1994) and the ability to change the heart rate in this split-second way is thought to be a sign of health, indicating that the body is able to constantly adjust to changing circumstances. So, at rest, high heart rate variability (HRV) is taken to be a sign of a fit, unstressed body. On the other hand, low or declining HRV (very constant intervals between heart beats) is taken as a sign that the body is stressed to the point that it is simply unable to make moment-to-moment adjustments to heart rate.

HVR has become very important for athletes and for people wanting to measure their fitness so that there are now countless devices for measuring it and apps for telling you how fit or stressed you are. HVR in humans has been linked to risk of cardiovascular disease and to psychological states such as anxiety and depression (Thayer et al. 2012). There is also a great deal of interest in fitting similar devices and applying the same ideas to non-human animals and using HRV as a measure of welfare (von Borell et al. 2007; Kovács et al. 2015). The technology has advanced to the stage where it is possible to fit a heart monitor on any animal from a cow to a fish and so get a continuous record of what its heart is doing as it moves around or is exposed to pleasant or aversive situations.

Decreased HRV has been detected in some animals in apparently aversive situations such as young calves being kept in social isolation (Clapp et al. 2015) and pigs exposed to a sudden noise (Leliveld et al. 2016), but the results are not consistent. Lameness is a serious health issue for dairy cows and it might be expected that lame cows would be 'stressed' and therefore show a decrease in HRV. However, cows that were lame showed an increase in HRV compared withcows that were not lame (Kovács et al. 2015), possibly because the lameness made them less active (Ede et al. 2019). While dogs that were separated from their caregivers (an apparently negative situation for them) showed decreased HRV, they also showed a decrease when being petted by their caregivers (assumed to be positive) (Katayama et al. 2016).

Some of these inconsistencies may be due to the exact way in which HRV is measured or the state of arousal the animal is in at the time when the measurements are made and so perhaps may be resolved by more sophisticated ways of analysing heart rate data (Zupan et al. 2016) but currently the value of HVR in welfare is still debatable (Kovács et al. 2015; Ede et al. 2019). Like other measures of stress, HRV can seem like the magic bullet for non-invasive welfare assessment but has the same problem of being very good at indicating that the animal is aroused or even just active, but much less good at indicating whether the animal is in a situation it regards as positive or negative. Current measures of HVR do not reliably indicate valence and so are not well correlated with what the animal wants.

Skin, eye and comb temperature

Another measure of arousal that has been misleadingly referred to as a measure of welfare is body temperature. When the body prepares for action, one of the changes that occurs is an increase in deep body temperature. Blood flow to the core is increased, diverting blood away from the body surface, which in turn leads to a drop in peripheral temperature—hyperthermia (Zethol et al. 1994). Techniques such as infra-red thermography now allow completely non-invasive measurements of tiny changes in the temperature of exposed parts of the body surface such as the eyes (Stewart et al. 2005; Gomez et al. 2018) or, in the case of chickens, the comb and wattles. For example, when chickens are handled, comb temperature shows a rapid drop of up to 2°C followed by an increase so that after 20 minutes skin temperature is actually higher than before the bird was picked up (Edgar et al. 2013; Herborn et al. 2015).

As with so many other measures of 'stress', however, very similar responses are seen in the presence of both positive and negative events. Thus dogs show a decrease in eye temperature when subjected to a veterinary examination (Travain et al. 2015) but a similar response when given treats (Travain et al. 2016), and cows showed a decrease in eye temperature when confined in a cattle crush to have their feet trimmed that was no different from the decrease that occurred when they were given highly palatable food (Gomez et al. 2018). Moe et al. (2012) trained chickens to associate a light being switched on with the reward of a mealworm (a treat highly prized by chickens) and found that the combination of light plus mealworm produced an average drop in comb temperature of 1.5°C. They concluded that the chickens' eye temperature response in this highly positively reinforcing situation was so similar to that reported for aversive situations such as handling that eye temperature was a very poor indicator of valence and much better seen as a measure of arousal.

Miscellaneous measures: shells and eye white

A variety of other responses to 'stress' have also been suggested as possible measures of welfare, but most of them turn out, on closer examination, to be better seen as measures of arousal rather than good or bad welfare because they lack that crucial element of valence. They tell us that the animal is aroused and ready for action but not whether it is aroused because it is in a positive emotional state and faced with a reward it wants, or aroused because it is in a negative emotional state and trying to avoid or escape from a punishment. For example, distortions of hens' egg shells, such as changes in shell thickness, were originally described as responses to stress as they occurred in response to environmental disturbances that were assumed to be negative to hens (Hughes et al. 1986; Solomon et al. 1987). However, similar shell distortions were also found in hens given access to an enriched environment, shown independently to be a preferred option (Dawkins et al. 2004). Similarly, the extent to which cows show the whites of their eyes was first described as a response to frustration in a cow unable to get to what it wanted (Sandem et al. 2002; Sandem and Braastad 2005) but it was subsequently shown that the amount of eye white a cow shows is similar in both positive and negative situations (Gomez et al. 2018). Once again, we have to be careful not to interpret an increase in arousal as a decline in welfare.

Conclusions

In this chapter, we have seen that linking animal welfare to physiological symptoms of 'stress' is not straightforward and has even been misleading. On the face of it, measuring physiological changes such as heart rate, body temperature and hormone secretion appears to be the best way of giving 'welfare' a firm scientific basis by invoking underlying mechanisms of physiology and immunology. However, describing these changes with the word 'stress' has seriously hampered the interpretation of physiological mechanisms in welfare terms.

Once again, animal welfare science has been, and indeed still is, bewitched by the words it uses. To be told that something invokes a 'stress' response in an animal immediately implies that the animal's welfare must have been compromised. 'Stress' has such strong emotional overtones of negativity, unpleasantness and poor welfare that it seduces us into thinking that if anything carries the label 'stress'—such as a 'stress hormone'—it must be indicative of negative valence, of an animal being in a situation that it does not want to be in. It implies unpleasantness, punishment and poor welfare.

In fact, the 'stress' label carries no such authority. Time and time again, we have seen that so-called measures of physiological stress are much more accurately seen as measures of arousal or preparation for action that can be similar whether the animal is finding something aversive or pleasurable. In other words, they completely fail to indicate valence, while giving the impression that they do. In that sense, invoking 'stress' has hampered progress in animal welfare science as it has blurred the line—totally crucial to animal welfare—between what animals find rewarding and what they try to avoid.

The physiological changes that animals show in response to a variety of circumstances are valuable contributors to the assessment of their welfare, but can be positively misleading if used on their own. As 'correlates' of welfare they often fail to distinguish situations that animals find pleasurable from those they find aversive and so should always be used in conjunction with independent measures of what animals want.

What is described here is only a brief outline of what is an immensely complicated series of reactions and feedbacks within the body. A very readable introduction to this complexity is given by Sapolsky (1994) and more detailed descriptions of the SAM and HPA responses are to be found in Mormède et al. (2007).

8

Animal Welfare with and without Consciousness

So far in this book, we have defined 'welfare' without any reference to consciousness, subjective feelings, sentience or indeed anything other than what we can see and measure objectively. We have seen that there can be a fully scientific study of animal welfare with clear methods, aims and standards of evidence that does not depend on having first solved the hard problem of consciousness and yet can still deliver valuable information on what makes animals healthy and what gives them what they want. In that sense, a definition of animal welfare does not need to include any element of consciousness. For many people, however, animal welfare without consciousness may be coherent and scientific but it is missing the most important component of all—namely, what animals *feel* (Dawkins 1990; Duncan 1993; Broom 2014; Mellor 2019). The subjective experiences of animals, particularly their capacity to feel pleasure and pain (sentience) is for many people what matters over and above what behaviour they are doing or what hormones are coursing through their veins. In this second sense, then, a definition of animal welfare very much does need consciousness and without it is incomplete and empty.

In this chapter, we will look at the arguments for and against adding conscious experiences to what we mean by animal welfare. Note that the issue being addressed here is not whether non-human animals have conscious experiences. The issue is whether conscious experiences of pleasure, pain and suffering should be *part of the definition* of what welfare is. These two issues are quite separate. It would be quite possible to believe that many animals have conscious experiences but at the same time to hold the view that, in the interests of objectivity and scientific testability, conscious experiences should not be included in the scientific definition of animal welfare. Equally, it would be quite possible to believe that sentience was beyond our current scientific understanding but that for pragmatic reasons (such as appealing to the public or getting legislation passed), sentience should be included in a definition of animal welfare on grounds of common sense or intuition.

Animal welfare defined with consciousness

Most current definitions of animal welfare contain either an explicit reference to conscious experiences of pleasure, pain and suffering or an implicit reference such as 'positive

The Science of Animal Welfare: Understanding What Animals Want. Marian Stamp Dawkins, Oxford University Press (2021).
 DOI: 10.1093/oso/9780198848981.003.0008

emotional state' or 'positive affect' (Fraser 2008; Proctor 2012; Broom 2014; Mellor 2019). Fraser (2008), for example, sees animal welfare as having three separate components—health, natural behaviour and 'affective state', by which he means conscious experiences of positive and negative emotions. Mellor (2019) also emphasizes the central role of conscious experiences in welfare assessment through the concept of 'welfare-aligned sentience' (p. 9), which he argues 'confers a capacity to consciously perceive negative and/or positive sensations, feelings, emotions or other subjective experiences which matter to the animal'. Many scientists accept that there may be doubt over consciousness in animals, but, if there is, animals should be given the benefit of it. This is in line with the view of many non-scientists that animal consciousness is so obvious that questioning it in any way is just unnecessary quibbling over words and, worse, detrimental to the welfare of the animals themselves. Birch (2017) has argued that consciousness should be part of the definition of animal welfare on the grounds that both politicians and the public are more likely to take animal welfare seriously if they are told the animals involved are sentient.

I have a great deal of sympathy with the view that a definition of welfare should be based on the ability to feel pleasure and pain. Indeed, I have a proven track record of saying that animals have conscious experiences at a time when such a view was controversial and unfashionable. In 1980, I published a book with the unequivocal title *Animal Suffering: the Science of Animal Welfare.* In 1990, I followed this up with a target article in *Behavioral and Brain Sciences* that began with the words 'Let us not mince words: Animal welfare involves the subjective feelings of animals.' This view was shared by numerous other scientists (Duncan 1993; Webster 1994; Broom 2014) and reiterated in a further book of mine on animal consciousness (Dawkins 1993). So why am I now suggesting that sentience should not form part of the definition of animal welfare? The reason is that the main problem confronting animal welfare science has changed.

As I saw it in 1980, the main issue then was to convince people that consciousness in animals needed to be taken seriously and that animal welfare should form the basis of a new and more scientific research programme than had been undertaken up to that time. I could see problems with studying consciousness in animals, of course (Dawkins 1990, 1993) but I thought they were not much greater than those of studying it in other people. We do not know for certain that other people are conscious in the same objectively observable way that we know that they have red blood or five fingers but we do not let that stand in the way of assuming that they are. They look and act like us and they tell us what they feel in ways that sound similar to what we feel. So we assume that other humans have conscious experiences like us, even though we cannot prove it.

Similar logic, I thought, could be applied to animals. We can study their behaviour and physiology objectively and, where these are sufficiently like our own, we can infer that they have conscious experiences like us too. An obvious difference is that we do not have words to ask animals what they are feeling but we do have other ways of 'asking' them to give us their point of view (Dawkins 1990, 1993). Absence of words was therefore not a major obstacle. The problem was to overcome the reluctance of scientists, widespread at the time, to admit that consciousness was a respectable subject for scientific study. Behaviorism (the view that only behaviour and physiology could be studied scientifically and consciousness could not) had a firm grip on both human psychology and the study of animal behaviour for much of the twentieth century and it was not until the latter part of that century that even the study of human consciousness really gained scientific credibility (Boakes 1984; Dawkins 2012). The beginning of acceptance of animal consciousness studies as legitimate science can be dated quite precisely to the publication of Donald Griffin's landmark book *The Question of Animal Awareness* in 1976. Griffin used the growing number of studies of animal intelligence, problem solving and the complexity of animal behaviour to argue that many species are consciously aware of what they were doing. And if they are aware of what they were doing, then it is a short step to saying that they are consciously aware of different emotions such as fear and pain and that a science of animal welfare should fully embrace the study of what animals feel.

The main problem today, however, is not persuading scientists or politicians to take animal consciousness seriously. That job has been, if not fully accomplished, then at least taken well in hand, as is evident from the legislation around the world that now makes specific reference to sentience (Mellor 2019) and is built into the European Union's Treaty of Lisbon (2009) and the Global Animal Welfare Strategy of the World Organization for Animal Health (OIE 2012). The problem is now persuading animal welfare scientists to be clearer about what they mean by 'good welfare'. The problem was stated by John Webster many years ago in these blunt words: 'Any proper discussion of animal welfare requires a comprehensive and unsentimental definition and analysis of welfare as perceived by the animals themselves. Single sentence definitions such as "Welfare defines the state of an animal as it attempts to cope with its environment" (Fraser and Broom 1990) tend to have the self-referential flavour of "A rose is a rose" and do not really advance our understanding' (Webster 1994, p. 10).

Webster's own solution was to convince the Farm Animal Welfare Council to adopt the Five Freedoms, which, because they were down to earth and practical, have had a huge influence around the world on the way people think about animal welfare. But even the Five Freedoms use some terms that are difficult to define, such as 'freedom', 'fear', 'distress' and 'normal behaviour', which leave it unclear how these are to be actually measured in practice. All I am suggesting is that the time has now come to be even more explicit about the steps that need to be taken to improve welfare by using even simpler terms, with an even clearer prescription of what evidence we need to collect, summed up in the phrase 'what animals want'. We ask the animals what it is they want (or do not want) and use that information to define what is good or bad for their welfare. The definition is then not circular or self-referential or dependent on terms that may mean different things to different people.

A clearer definition of animal welfare would provide support for the increasing number of farmers, zoos and other people who are willing to make major changes to improve the welfare of their animals but who also want evidence as to what those changes should be. They are no longer satisfied with being told by well-meaning people how to improve animal welfare on the basis of intuition or what, from a human perspective, seems like a good idea. They want animal-centred evidence as to what actually improves welfare in practice.

We have thus reached what could be called the second stage in attitudes to animal welfare. The first stage was when concern for animal welfare was a minority view and the public had to be convinced that it was important. The second stage is now, when concern for animal welfare is widely accepted as an ethical good for society and where the main concern has become implementation—actually making the major changes that are being called for (Grandin 2019). All too often the evidence is not there and one of the obstacles to obtaining it is not having a clear definition of what animal welfare is. Agreeing on a definition will be crucial to making this second stage a time of real improvement in the welfare of animals.

The benefits of including consciousness in that definition depend on how well we now understand consciousness itself. If science can now tell us a great deal about consciousness in ourselves and in other animals, then it will be relatively easy to agree on a definition that includes a reference to conscious experiences in animals and that could benefit animals by including it. But if science is still a long way away from being able to give a full account of consciousness, then trying to make it the basis of the definition of welfare will only add to the confusion surrounding an already difficult to define term and make animal welfare scientists look as though they are unclear about what they mean. Animals will not benefit from such a lack of clarity. A definition that contains a major unknown will itself be an even greater unknown and make it more difficult to implement change in the lives of animals. So how well does science now understand consciousness?

Human consciousness and animal consciousness

A promising starting point for seeing how well we currently understand consciousness is to look at the one species we know for certain does have conscious experiences—our own. With human consciousness we at least know what it is from our own personal experience and we also have the advantage of being able to use language to ask other people what their conscious experiences are like. If we could identify the characteristic 'signature' of consciousness in our own brains, then it might be possible to look for similar signatures in the brains of animals that cannot tell us in words what they are feeling. Even a theory of consciousness that did not explain how conscious experiences arises from brain tissue (i.e. that did not solve the 'hard problem' we discussed in Chapter 2) might at least indicate what brain structures or types of brain activity were correlated with consciousness. We could then see whether similar brain structures or activities are also found in other species. Insights from human consciousness could therefore become our way of coming to grips with animal consciousness.

The past 25 years have seen an intensive search for what have become known as neural correlates of consciousness or NCCs in humans (Block 1995; Tononi and Koch 2015), a search that has been greatly helped by a variety of new techniques. Chief among these is brain imaging that has effectively made the human skull transparent and allowed scientists to see what a person's brain is doing while they are thinking, remembering, responding to stimuli or doing a variety of tasks (Owen 2013; Dehaene 2014). Functional magnetic resonance imaging (fMRI), for instance, shows which parts of

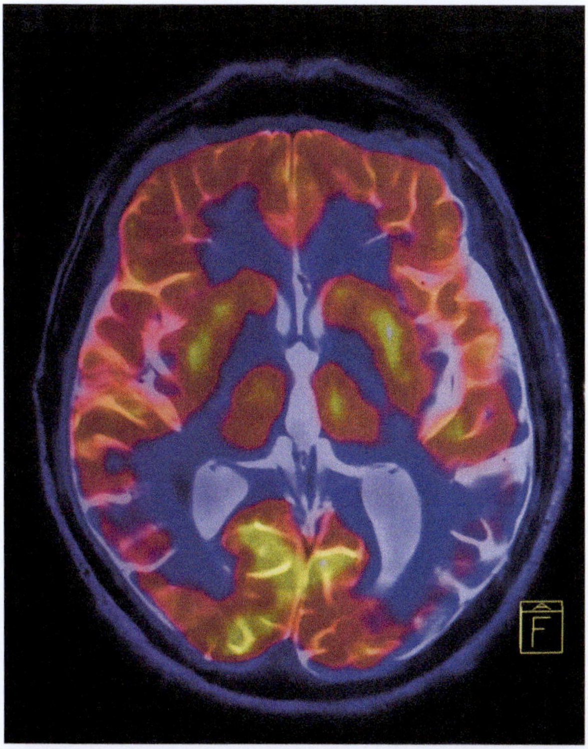

Figure 8.1 *The human brain revealed. Functional magnetic resonance imaging (fMRI) makes use of the fact that when neuronal activity increases in a particular part of the brain, the glial cells surrounding the active neurons open up nearby arteries, so that 2 3 seconds later, the blood flow to the neurons increases, bringing with it haemoglobin molecules carrying oxygen. As the haemoglobin molecules give up their oxygen to the active synapses and neurons, they become slightly magnetic. This change in the magnetic characteristics of blood in the presence of active nervous tissue can be picked up outside the skull by sending in radio waves, listening to the returning echoes and then using these to construct a colour map of whereabouts in the brain the most obvious magnetic changes are happening. Photo credit Semme (Shutterstock).*

a living, thinking brain are most active at any one time by detecting changes in blood flow. The more active neurons are, the more oxygen they need and the more blood is diverted to them and this can be shown as a map in which areas of high neural activity appear as different colours (Figure 8.1). Scientists can then see the changing patterns of brain activity reflected on the map as patterns of changing patches of colour (Figure 8.1).

fMRI does not detect neuronal activity itself. This can change rapidly within a few milliseconds but what fMRI can do is to pick up gross changes in the blood supply that happen a few seconds later. So what it is doing is detecting whereabouts in the brain the greatest activity was a short time earlier. It is also a very blunt instrument—it can only measure the changes that occur in groups of a million or so neurons at once. Nevertheless, it has proved a powerful way of exploring human consciousness through highlighting which parts of the brain are most active when people report on different kinds of conscious experiences (Dehaene 2014; Koch 2017).

fMRI becomes a particularly useful way of studying human consciousness when it is combined with a technique known as 'masking', which makes it possible to move between conscious and unconscious processing of incoming visual information. Exploring what the human brain does as it switches between unconsciousness and consciousness is obviously a step forward in understanding the nature of consciousness itself. Masking consist of presenting images such as photographs of faces for such a short time (less than 30–40 milliseconds) that people are not aware of having seen anything at all. It is called 'masking' because the very short exposure of an image is achieved by presenting another image or mask both before and after it. The image of interest is thus framed in time by the mask, giving tight control over the exact length of time for which the image itself is visible (Dehaene 2014). Images presented for relatively long periods (more than a fifth of a second) are generally seen and reported on, whereas masked images presented for less than 30 milliseconds are not seen or reported on.

An extraordinary result of such masking experiments is that, although people can sometimes be consciously completely unaware of having seen the briefly presented images, their behaviour shows that they have in some sense 'seen' them, although unconsciously. For example, if they are shown photographs of human faces with happy, sad or neutral expressions for less than 30 milliseconds, they report that they see no faces, but at the same time, their own facial muscles are mimicking the expressions of the faces they supposedly cannot see (Dimberg et al. 2000). Photographs of happy, angry or sad faces presented too briefly for people to be consciously aware of them also affect their emotional responses to neutral stimuli so that the same shape is perceived as optimistic or pessimistic depending on the expression on the 'unseen' face (Murphy and Zajonc 1993; Sato and Aoki 2006; Axelrod et al. 2015). The explanation for this appears to be that there are two different ways of seeing—one route that enters consciousness and another route that remains unconscious but can still affect behaviour.

Further evidence for two different ways of seeing has also been found in human patients who have suffered a particular kind of damage to their visual cortex where incoming visual processing takes place. Such patients have normal eyes but damage higher up the visual pathway that leaves them with a large blind area of their visual field where they say they cannot see anything at all. Weiskrantz (1997) started to investigate exactly what these patients could see by deliberately putting objects in their blind field. To his surprise, he found that, although the patients kept telling him they could not see anything, they were often able to perform visual tasks such as reaching out and touching an object or correctly orienting it to be put through a vertical or horizontal slit. They also gave the right answer well above chance when asked to guess what was in their blind field, and then were astonished when told that they were correct. But even as they carried out these tasks, perfectly competently, they would insist that they could not see anything at all and even became quite annoyed when first asked to guess what was in front of them on the grounds that, as far as they could see, there was nothing there at all.

Weiskrantz (1997) called this condition 'Blindsight' because his patients said they could not consciously see but at the same time were correctly able to perform visually guided behaviour. They were both blind (consciously) and sighted (unconsciously). Blindsight patients have subsequently been shown to respond emotionally to visual images without being conscious of what they are responding to. When shown photographs of human faces with different expressions they say that they cannot see any faces at all, let alone describe what emotion they are expressing, but at the same time their own facial expressions mimic those on the faces they insist they cannot see (Tamietto et al. 2009; Brown et al. 2019).

From these and many other studies, it is now increasingly clear that the search for human consciousness has led to an unexpected conclusion. This is that, greatly through consciousness seems to dominate our own human existence, much human behaviour is in fact carried out quite unconsciously (Axelrod et al. 2015; Rolls 2020; Cleeremens et al. 2020). Many complex tasks that may appear to have the hallmark of consciousness can be done without it, including reading and doing arithmetic (Sklar et al. 2012; Schelonka et al. 2017), driving a car and even learning the rules of a complex game (Colman et al. 2010).

Dehaene (2014) expresses this in quite extreme terms by saying that for most of the time most of the operations of the human brain are carried out without consciousness at all. 'A wild profusion of unconscious processes weaves the textures of who we are and how we act', is how he puts it (p. 191). He argues that, even for us, conscious experiences are rare and only a few brain operations make it to conscious awareness. Any theory of consciousness worth having therefore has to be able to explain how we can do so much unconsciously and also what happens as we change between conscious and unconscious states, particularly if we want to use the same theory to decide which non-human species have conscious experiences.

Now that we know that so much complex human behaviour can be performed without consciousness, the question of animal consciousness obviously becomes even more difficult. It is not sufficient to equate consciousness in a vague, general way with complexity of behaviour because complex behaviour can be performed, even by us humans, without consciousness (Shettleworth 2010). There needs to be a more explicit and testable theory of when there is and when there is not conscious awareness (Carruthers 2019).

Can theories of human consciousness tell us about animal consciousness?

Theories of human consciousness abound, none of them conclusive and many of them overlapping. Despite the huge research effort that has been put in and the breakthrough techniques that are now available, despite numerous journals and countless books dedicated to understanding consciousness, it is still possible to look at everything that has been done and conclude that we are as far away as ever from understanding the biological basis of consciousness (Cleeremans et al. 2020). Nevertheless, much progress has been made and it is now possible to discern several different types of theories (Carruthers 2019; Michel et al. 2019), many of which are directly relevant to the issue of consciousness in animals.

The first kind of relevant theory looks for consciousness in particular brain structures. The appeal of this approach is that if we could identify the anatomical site of conscious awareness in ourselves, we could then look for similar consciousness structures in other species. Finding them would then be a very strong indication that they too were conscious.

Several different parts of the brain have been suggested as the seat of human consciousness. One of these is the upper brain stem, partly on the grounds that this is where motivational and sensory information is integrated before behaviour occurs (Merker 2007) and partly also because damage to one particular part of the brain stem has been found with people being unconscious in a coma (Fischer et al. 2016). The part of the brain stem known as the reticular formation is associated with the transition from sleep to waking, suggesting that it might have a role in consciousness itself.

Rather clearer evidence implicates the cerebral cortex, particularly the fronto-parietal areas, as essential to conscious experiences (Dehaene and Changeux 2011). For example, using the 'masking' technique we discussed earlier, Grill-Spector et al. (2000) showed pictures to people for different lengths of time while at the same time scanning parts of their brains for activity and asking them what they could consciously see. She found that the primary visual cortex (the early part of the visual pathway in the brain) was activated by the sight of all the pictures, regardless of how long they were shown and whether people were conscious of having seen them or not. However, further up in the visual pathway, in the fusiform gyrus and lateral occipito-temporal areas of the cortex where processing of faces, words and objects takes place, not all pictures resulted in brain activity. Only images presented for long enough for people to report that they consciously saw them were correlated with activity in these cortical areas. Activity here thus seemed to be specifically associated with people consciously seeing the images in front of them.

However, 'being associated with' is not the same as being the seat or origin of consciousness. Eyes are essential to conscious vision but that does not mean that eyes give rise to consciousness. In any case, conscious awareness may not reside in one specific location as different parts of the brain seem to be associated with being conscious depending on what someone is doing. For example, the left parietal and prefrontal cortex are involved in the conscious perception of words (Dehaene et al. 2001), while the amygdala and the insula appear to be involved in the conscious perception of faces (Phillips et al. 2004).

Studies of the brains of people waking up after anaesthesia also show that there is no one brain area that uniquely indicates when they have regained consciousness. Coming round from most kinds of anaesthetics does not seem to involve any one specific part of the brain, but rather a general, but little understood change in the way in which cells all over the brain communicate with each other (Kulli and Koch 1991). So even when we experience what seems like a clear and sudden transition from oblivion to full consciousness, we cannot pinpoint a change in a specific brain area.

As we saw earlier, brain imaging studies, particularly when combined with masking experiments, have told us a great deal about which areas of our brains are active at different times and that there are indeed correlations between brain activity and whether people report that they are conscious of something (Dehaene 2014; Blackmore and Troscianko 2018) but they have also told us that there is no one place in the brain that is uniquely associated with consciousness. Anatomically, consciousness remains elusive. It can be associated with generalized effects in many parts of the brain as in anaesthesia or it can be associated with activity in a particular area, but even then, exactly which area can vary depending on what the brain is doing at any one time. So far, no-one has found the one and only seat of consciousness in the brain or any special consciousness-related nerve cells, or any unique structure we can identify as being necessary for awareness (Boly et al. 2013; Stoerig 2007; Koch et al. 2016; Feinberg and Mallat 2019). Not being able to tie down consciousness to a specific anatomical location thus complicates the search for consciousness in other species because it gives us so little idea of what to look for.

So a second kind of theory—and one particularly relevant to animals with brains very different from ours—has assumed importance in consciousness studies. This approach involves trying to understand what a conscious brain is doing, rather than exactly how it is constructed. In other words, it shifts the goal of consciousness research away from anatomical details to the identification of the neural interactions associated with conscious experiences. This makes it well suited to the search for consciousness in animals such as birds, octopuses or insects. These animals clearly do not have the same brain hardware as we do but that logically does not rule out consciousness that is supported by different anatomies (Barron and Klein 2016; Tye 2017). After all, flight happens with different anatomies so why not consciousness? Birds, bats and insects have evolved very different sorts of wings but they can all still fly. And if consciousness can similarly arise in very different sorts of brains, then anatomy becomes less important than specifying what kinds of neural processes might be advanced enough to suggest consciousness (Barron and Klein 2016; Bronfman and Ginsberg 2016; Feinberg and Mallatt 2019).

There is now a large number of theories all of which claim to have found the link between neuronal activity and consciousness (Boly et al. 2013; Feinberg and Mallatt 2019; Bronfman and Ginsberg 2016). One idea that many of these theories have in common is that consciousness involves integrating data from many different parts of the brain, while unconscious processing is much more localized. For example, the global workspace theory (Baars 1988, 2005; Baars et al. 2013; Dehaene 2014) sees consciousness as a 'theatre' in which conscious sensory and other information is 'broadcast' widely across the brain to an 'audience' of expert brain networks. These local brain networks are, individually, unconscious but as they broadcast and share information globally across the brain, this gives rise to conscious experience. Global workspace theory thus explains the fact that the brain can operate either unconsciously or consciously as an ability to switch between local and global processing. It also provides an explanation about what is special about the conscious mode of operation—it is when information from different parts of the brain are brought together to give insights not available to individual areas.

When we make decisions, for example, we use information from different parts of the brain. If all we knew was the colour of an object, we might not be able to recognize it, but bringing together different sorts of sensory information such as its colour, shape and smell makes it much easier to decide what it is (Tononi 2008). Coordinated movement of the body also needs sensory and motor information of many different sorts (Barron and Klein 2016). Since many animals, including insects and crustaceans, have what seem to be a similar ability to integrate different sorts of information, it has been suggested that they too are conscious (Barron and Klein 2016). However, there is still no agreement as to exactly what 'broadcasting' means or how much 'integration' is needed to move from unconscious gathering of information to conscious perception (Boly et al. 2013; Carruthers 2019). So although considerable excitement has been generated by these integrative processing theories, their lack of precision means that they have confused rather than clarified the search for consciousness in animals, making it possible to claim that every animal that shows any degree of linkage between different parts of its nervous system could be claimed to be conscious.

A third group of theories concentrates on what a brain can achieve when it is conscious as opposed to what it can do when it is operating unconsciously. The search has been for some sort of discontinuity in cognitive or other ability—some task that can only be done with conscious awareness. There are many such proposed boundaries and one of the most widely used is one that we came across in Chapter 4—that between fixed innate responses and irreversible habits on the one hand and goal-directed behaviour that involves completely arbitrary responses on the other (Balleine and Dickinson 1998; Dickinson 2012; Pennartz et al. 2019).

Of course, as we have seen all along, a significant boundary in behavioural achievement, does not say anything at all about whether it is also the boundary between unconscious and conscious processing but, from a behavioural point of view, it really does seem to represent a step change in what an animal is capable of learning. Innate preferences and reflex responses can be pre-programmed by the genes. If the animal learns anything, it learns simple fixed habits that are difficult to reverse. But with goal-directed behaviour, the degree of pre-preparedness is much less and the animal has to learn for itself which of the many possible behaviours it could do is the one that results in the reward or the punishment (Balleine and Dickinson 1998; Dickinson 2012; Pennartz et al. 2019). Genetics is still involved in that genes specify what the animal finds pleasurable or aversive, and provide the capacity to learn, but the genetic specification is much more open-ended. What behaviour the animal has to do to obtain pleasure or avoid punishment is not fixed in advance, giving it great flexibility in adapting to novel situations and the ability to change as its environment changes. In evaluating what animals want, we saw in Chapter 4 that this ability to 'try anything' to obtain a reinforcement does indeed represent a step change in learning ability. However, whether it is also a marker for the transition from non-sentience to minimal consciousness as Bronfman and Ginsberg (2016), among others, have claimed, is a separate issue. Without a theory of what consciousness is, it is not possible to equate goal-directed behaviour plus reinforcement learning with conscious awareness. It is an assertion but nothing more.

Another point where it has been proposed that consciousness 'kicks in' is with the ability to have higher order thoughts or HOTs (Rosenthal 1993, 2005). HOTs are 'thoughts about thoughts' or the capacity to think about your own thoughts; for example, to be able to reflect on a planned course of action and decide that your proposed plan would not work. 'I am looking at food' is a first order thought. 'I have a thought that I am looking at food' is a second order thought and 'I wonder if my thought that I am looking at food is correct' is an even higher order thought. HOTs clearly demand considerable cognitive ability, but their relationship with consciousness is unclear. Even HOTs theorists who believe that HOTs are necessary for consciousness, concede that not all HOTs are conscious. Machines can easily appear to 'monitor their own thoughts' by keeping a simple tally of the strength of their memories or the certainty of their predictions and acting accordingly (Insabato et al. 2010).

Rolls (2020) has recently suggested that it is only a specific sort of HOT—higher order syntactical thoughts or HOSTs—that reach consciousness. HOSTs involve syntactical processing, which is the ability to mentally represent the world symbolically and then manipulate those symbols in different ways. This might happen, for example, when we think about a plan of action that we have just made, realize that it is flawed, identify which step in our plan is causing the problem and then work out how to correct it—all in our heads. In us, this symbol manipulation is done with verbal language so HOTs theories are sometimes seen as ruling out consciousness in non-language-using animals (Blackmore and Troscianko 2018). However, verbal language is only one way in which mental representations of the world can be manipulated, so HOST theory, with its emphasis on syntax, does not preclude consciousness in animals even in those that are unable to use words to report on their experiences. Instead it makes the entry point into consciousness the ability for syntactical reasoning and planning with the explicit prediction that multi-stage, fault-correcting planning cannot be done unconsciously (Rolls 2020). A hypothesis with predictions that can be tested is a step forward, even though the specific tests have yet to be carried out.

There is thus a great diversity of ideas about human consciousness and no agreed theory of what consciousness is or what gives rise to it. Far from clarifying consciousness in animals, the study of human consciousness has therefore only shown us how much we still do not understand about consciousness. It has shown how much we humans do without consciousness and so has left us more divided than ever about what to look for in other animals. Claims that 'with this brain structure' or 'with that ability' animals went from unconscious behaviour to conscious wants are often highly plausible, particularly when there seems to be some functional difference between those with and without the structure or ability in question. However, they remain inferential and open to challenge (Evers and Sigman 2013; Gutfreund 2018). For every such claim, someone else can make a counterclaim and there is, as of now, no satisfactory way of deciding who is right. All that it is possible to say with certainty is that during the course of the billions of years that life has been evolving on Earth, there was an increase in the complexity in the way organisms obtain what they want. There have been some identifiable milestones along the way, such as being able to actively move towards or away, simple learning, more flexible operant learning directed by goals, and ultimately thinking and planning ahead. But many of these steps, including operant conditioning, can take place without consciousness, so exactly when or how or even why consciousness became part of the picture is still a matter of speculation (Boly et al. 2013; Dawkins 2015; Blackmore and Troscianko 2018).

Consciousness remains the hardest problem in biology. The study of human consciousness, far from being the means of clarifying what consciousness is, turns out to confront us with even greater problems because we humans can do so much without it. We do not know whether other animals are more like us when we are conscious or more like us when we are behaving unconsciously. We have yet to identify any task that can only be performed by a conscious being. We cannot point to any specific brain structure uniquely associated with conscious experiences. Our knowledge of human consciousness is therefore of limited value in illuminating consciousness in other species because we understand it so little in ourselves that we have no clear idea of what to look for in other species.

The result of this lack of certainty is the extraordinary situation that we discussed in Chapter 2 and that has been the inspiration behind this book—that somebody, somewhere has proposed, on apparently reasonable grounds, every single possibility about which animals are conscious. Depending on which discontinuity or continuity you point to, you can make the case for only humans, only primates, only mammals and birds, or all fish, insects, crustaceans and others being members of the 'consciousness club' (Boly et al. 2013).

The philosopher Dan Dennett provide a stark summary of our state of knowledge of consciousness. Having pointed out that other great mysteries such as the origin of life and the evolution of different species have been, if not solved, then at least tamed to the point that we know how to think about them, he writes: 'With consciousness, however, we are still in a terrible muddle. Consciousness stands alone today as a topic that often leaves the most sophisticated thinkers tongue-tied and confused' (Dennett 1991, p. 22). We do not need muddle and confusion in the definition of animal welfare.

Animal welfare defined without consciousness

As has been repeatedly emphasized, leaving consciousness out of the definition of animal welfare does not imply that animals are without conscious feelings. It is simply a statement that animal welfare science is more rigorous, more scientific and therefore in a better position to promote animal welfare if it does not *define* animal welfare to include something as elusive, confusing and difficult to define as consciousness.

A hypothetical example will show how invoking consciousness or sentience confuses rather than clarifies. Suppose someone proposed X as a 'measure' of welfare. 'X' could be a hormone, a particular behaviour, eye temperature or any of the other measures we have discussed in this book. X is accordingly validated by asking whether it is indicative of improved/decreasing health and also whether it is indicative of animals having what they want or want to avoid. A barrage of tests shows that, in practice, it does not perform very well in any of these tests. Its connection with long-term health cannot be established and it appears to be a measure of arousal rather than valence. It is clearly not a good correlate of either health or what animals want and so, on this definition, X is not a good measure of welfare and should not be relied on in welfare assessment.

Now see what happens if we try adding consciousness to the definition, so that the validation includes, in addition to the first two criteria, asking whether it is also a good measure of what the animal is feeling. Since we do not know how to assess conscious feelings independently of the other two criteria, X cannot be dismissed so lightly. Maybe, some might argue, it does reflect what an animal is feeling, even if it does not indicate either heath or what the animal wants. Perhaps it captures that magical 'extra' ingredient. If we don't know what it is, we can't be sure that it isn't there. X therefore remains on the list of valid welfare indicators. If X is also described by some pejorative name such as a 'stress' hormone or 'optimism' or a 'happiness response', then it is even more likely that X will be used as a welfare 'indicator', despite its demonstrable lack of ability to deliver information about either health or valence. The possibility that it might reflect conscious feelings trumps its other shortcomings.

Seeing consciousness as additional to and somehow superior to health and what animals want has therefore led to a lack of clear thinking among animal welfare scientists about what welfare is and what are the best ways to measure it. Restricting the definition to health and what animals want focuses minds on what is actually measurable, which 'measures' are most valid—and why.

Animal welfare defined with consciousness divides people. It pits animal welfare scientists against each other, with some claiming that animal sentience is an established fact and others saying it is likely but we cannot claim it as fact. Worse, it encourages confusion of thought through the deliberate use of ambiguous terms such as 'affect' and 'fear', which can be variously used either to imply that consciousness is involved or that it is not involved, depending on what audience the speaker is addressing. The result, as Le Doux (2014) has pointed out, is that people confuse themselves about whether they do or do not imply conscious experiences. Trying to clarify one term with no agreed definition (animal welfare) by making it dependent on yet another term with no agreed definition (consciousness) only creates confusion inside animal welfare science and risks giving the outside world an excuse for not taking its findings seriously. A weak, confused and ambiguous definition of animal welfare harms animals because it makes animal welfare science look weak, confused and ambiguous.

Animal welfare defined without consciousness, on the other hand, unites people with very different views about animals, both scientists and non-scientists. It provides a common ground for agreement. Everyone—even those who do not believe that any non-human animals are sentient or that even if they are, humans matter much more—can see the point of keeping animals healthy and satisfied. Animal health has such a major impact on human health through the food we eat, the likelihood of our catching animal-borne diseases and the sustainability of the human of life that even someone who cared little or nothing about animals themselves has to admit that animal health is important to human health and well-being (Dawkins 2012). 'Health and what animals want' can be used by everybody—including those who believe that animal consciousness has already been proved (e.g. Bekoff 2007), those who think that it is unproven but highly plausible, as well as those who think that our knowledge of consciousness is still so rudimentary that trying to include it is, as yet, beyond science.

Everyone—whatever their views on animal consciousness—can agree that making sure that animals are healthy and have what they want constitutes an important way of ensuring their welfare. And everyone—whatever else they choose to add to the definition of animal welfare—can agree that, in practical terms, satisfying these two criteria is the key to giving animals a high-quality life.

Animal welfare defined without consciousness does not rule out the strong possibility that many non-human animals have some form of consciousness. Nor does it say that consciousness is somehow unknowable and therefore forever beyond the reach of scientific enquiry. Nor does it even say that animal welfare scientists should not be studying sentience. On the contrary, it frees them to admit how little we currently know and to take part in one of biology's greatest adventures—trying to understand how nervous tissue gives rise to conscious awareness. Indeed, animal welfare scientists, with their research emphasis on affect, learning and decision-making in animals, are particularly well placed to make a positive contribution to the study of consciousness (e.g. Mendl et al. 2010). All that leaving consciousness out of the definition of animal welfare says is that, with our current state of knowledge, animal welfare should not be defined with respect to conscious experiences. The most useful pragmatic definition of animal welfare does not need consciousness and, more importantly, is better off without it.

Conclusions

The difficulty of studying consciousness, even in humans, is a major reason for the current lack of a universally agreed definition of animal welfare. Claiming (or giving the impression of claiming) to have solved 'the hard problem' by including consciousness in the definition makes animal welfare science look unscientific. Animal welfare science that is seen as unscientific will not be taken seriously by the outside world and will damage the prospects for improving the welfare of animals in the long term. Defining animal welfare without consciousness, on the other hand, provides a way of uniting people with very different views about animal welfare and focuses attention on what evidence needs to be collected to bring about genuine improvements in the welfare of animals.

9

Conclusions: A Universally Agreed Definition of Animal Welfare?

One of the main aims of this book was to arrive at a definition of animal welfare that was clear, made it obvious what evidence had to be collected to achieve good welfare, and could also be understood and agreed to by scientists and non-scientists alike. Without a clear definition, the whole idea of 'welfare' looks vague and unscientific. Without stating exactly what evidence needs to be collected to demonstrate good welfare in practice, it is impossible to judge whether or not an apparent 'improvement' really does improve the lives of animals. And without universal agreement, there will be constant arguments among politicians, farmers, animal charities, scientists and the public about whether this or that system is better or worse for animal welfare. An agreed definition, on the other hand, would provide a united front on what needs to be done and therefore increase the likelihood that improvements in animal welfare will actually occur. So has this aim been achieved? Does 'health and what animal want' provide the basis for such a definition? To answer this question, we need to first to look at what a universally agreed definition of welfare needs to do and what criteria it has to satisfy.

Test one: does it unite existing definitions?

A definition of animal welfare with aspirations to be universally adopted would be unlikely to gain general acceptance if it departed radically from definitions currently used by animal welfare scientists or ignored the wisdom and insights accumulated over the years. At first sight, this seems a tall order. Existing definitions appear to be extremely diverse, ranging from an emphasis on health and productivity (McGlone 1993), fitness and reproductive success (Broom 1998) longevity (Hurnik 1993), 'coping' (Broom 1998) to natural behaviour (Bracke and Hopster 2006), mental and physical health, positive affective state (Mendl et al. 2010) and combinations of many different measures such as those found in the Five Freedoms (FAWC 2009). As this book has shown, however, looking more closely at these different definitions already reveals a surprising measure of

The Science of Animal Welfare: Understanding What Animals Want. Marian Stamp Dawkins, Oxford University Press (2021).
 DOI: 10.1093/oso/9780198848981.003.0009

agreement. There is in fact a consensus that good welfare should involve high standards of animal health and also what amounts to a near consensus that it should involve animals having what they want, although not everyone would perhaps express it in quite these words. The extent of agreement that already exists is much greater than might appear. For example, the widely used Five Freedoms (FAWC 2009) are:

1. Freedom from hunger and thirst.
2. Freedom from discomfort.
3. Freedom from pain, injury and disease.
4. Freedom to express normal behaviour.
5. Freedom from fear and distress.

The first three of these Five Freedoms are about maintaining animal health, while the last two are about giving animals what they want (Freedom 4) or avoiding keeping them in situations they find aversive (Freedom 5). The same emphasis on the same two components is to be found in the results of a European-wide project, Welfare Quality® (2018), which defines good welfare as good housing, good feeding, good health and appropriate behaviour or, more specifically, as 12 requirements. These are:

1. Animals should not suffer from prolonged hunger.
2. Animals should not suffer from prolonged thirst.
3. Animals should have comfort around resting.
4. Animals should have thermal comfort.
5. Animals should have enough space to move around freely.
6. Animals should be free of physical injuries.
7. Animals should be free of disease.
8. Animals should not suffer pain.
9. Animals should be able to express normal, non-harmful social behaviours.
10. Animals should be able to express other normal behaviours.
11. Animals should be handled well in all situations.
12. Negative emotions such as fear, distress, frustration or apathy should be avoided whereas positive emotions such as security or contentment should be promoted.

Again, looking down this list, we can see that each of these requirements is promoting animal health, giving animals what they want or ensuring that animals are not kept in situations they demonstrably find aversive. The Welfare Quality® project emphasized the importance of defining welfare from the animal's point of view, which of course involves finding out what it does or does not want (Dawkins 1990).

In other words, health and what animals want are not concepts new to animal welfare science. They are restatements of what most people already mean by it anyway, but with the great advantage that they are expressed in down-to-earth language that tells us what we need to measure. 'Freedom' is an inspiring phrase but can leave us wondering what it means and how to achieve it. 'Normal behaviour' has, as we saw in Chapter 6, problems in defining exactly what it is. But translate both of these concepts into a demand for evidence about what keeps animals healthy and what they show us they want, and the uncertainty disappears. We can see what needs to be done to achieve them. So using these two criteria does not just fit in with other existing view of animal welfare; it reinforces them and gives them substance. It is not a new or alien definition. It just makes explicit what many people have been saying all along.

Test two: does it indicate which 'measures' of welfare are valid?

We have seen throughout this book that there is a vast array of different physiological and behavioural signs that have been suggested as 'measures' of welfare but considerable disagreement about which ones are most valid. These disagreements potentially have a negative impact on progress because lack of clarity in how to measure welfare makes it look as though the concept itself is vague and this in turn makes it difficult to make the case for what is needed to improve it. We have seen, however, that many of these controversies can be resolved by asking whether a suggested 'measure' is either a sign of good health or an indication that the animal has what it wants.

This has become increasingly important recently because many of the widely used measures of welfare, such as 'stress' hormones, changes in skin temperature or increase in activity, are now known be more accurately described as indications that an animal is aroused or active than signs of poor welfare (Chapters 5 and 7). Many behaviours and physiological changes turn out to be disconcertingly similar whether the animal is actively pursuing something pleasurable or it is trying to avoid something it finds aversive. Such measures of arousal need to be given 'valence' if they are to be considered reliable indicators of welfare. Now 'valence' is a technical sounding term for whether something is a positive or negative emotion (Mendl et al. 2010). But all it really means is that an animal is in a state it wants to be in (positive valence) or that it does not want to be in (negative valence). 'Health and what animals want' provides the valence for many of the proposed measures of welfare and anchors them firmly in an animal-centred view of the world. By using the words 'what animals want' and laying out how to determine what they want (Chapters 4 and 5), we have a way of establishing the valence that is such a crucial, but often missing, part of welfare assessment. We can avoid being distracted by emotionally loaded words such as 'stress' by going straight to the assessment of a situation from an animal's point of view. It then becomes clear that an animal can be 'stressed' by pleasurable situations as well as by aversive ones (Chapter 7) and what we need to give a balanced welfare interpretation is the animal's own verdict on a situation. 'Health and what animals want' thus passes this second test, too, by providing the valence that so many proposed 'measures' of welfare lack.

Test three: does it specify what evidence needs to be collected to demonstrate an improvement in welfare?

This is where 'health and what animals want' really comes into its own. The definition itself lays out clearly what evidence is needed to answer questions such as 'Are perches necessary for the welfare of broiler chickens?' or 'Do visitors adversely affect the welfare of zoo animals?' In the case of chickens, we need empirical evidence on what happens to the health of chickens with and without perches; we also need to know whether and how much the birds actually use perches. In the case of zoo animals, it may be more difficult, although not impossible, to carry out a full-scale health comparison between members of a particular species kept apart from visitors and the same species regularly visited by people, but it would certainly be possible to give those animals a choice between approaching visitors and somewhere to hide to get away from them and to see what they choose to do.

Collecting such empirical evidence moves us away from the temptation of trying to assess welfare by what we humans would like or what we imagine animals would like. Knowing what, in practice, improves animal health and what, from having asked them, they want is particularly important where some proposed change has important economic consequences or is likely to meet opposition. For example, if a farmer has to weigh up a loss in production (e.g. keeping fewer animals in a given space) against what are claimed to be improvements in the welfare of the animals, he or she will want actual evidence to support this claim. Many farmers are as concerned as anyone about the welfare of their animals but they increasingly demand evidence that the changes they are being asked to make really do matter to the animals themselves and are not just to make well-meaning consumers happy. Clear evidence on the health consequences of, say, keeping animals at different stocking densities can form the basis of reasonable discussions between people with very different views on a whole range of animal issues. Additional evidence on what the animals themselves want and how they respond when stocking density is changed can then shift emotional accusations of people not caring enough about animal welfare to more constructive discussions about how best to give them what they want.

Test four: is it understandable by non-scientists, including politicians and law-makers?

Here again 'health and what animals want' scores highly as a definition. Everyone can understand what it means and everyone can see whether the relevant evidence has been collected. Not everyone knows what 'valence' or 'affective state' means or why corticosteroid hormones are relied on so much by scientists, but everyone can understand what health is and what it is for an animal to want something it does not have or to want to get away from something it cannot escape. By being such a simple and down-to-earth formulation of what animal welfare is, 'health and what animals want' provides the common ground where people with apparently differing views can come together and begin to agree.

On these four criteria, then, 'health and what animals want' does well as a universally agreed definition of animal welfare. It arises naturally from existing definitions, it provides a way of judging the validity of different proposed 'measures' of welfare and it points directly to what evidence needs to be collected to improve the lives of animals. It has the further advantage that it can be easily understood by people both inside and outside science and so can facilitate dialogue between groups that may have very different ideas about animal welfare. It is fit for the purpose of being a simple, universally agreed definition of exactly what evidence counts as good welfare or poor welfare.

The welfare of non-human animals faces great challenges in our rapidly changing world. The need to feed the estimated 9 billion people who are estimated will be alive in 2050 (Godfray et al. 2010) and to meet demands for land, water and other resources has already led to conflicts about how to balance the welfare of humans with that of other animals (Dawkins 2012). To make sure that animal welfare is properly represented on the world's agenda when these matters are decided, we must avoid a definition that is muddled and confused. We need a definition of animal welfare that is clear to everyone and that will help us to find practical ways of improving the lives of animals.

'Health and what animals want' does all that. Adding a requirement that we also need to include what animals are feeling in the definition detracts from its simplicity, its clarity and its power to improve animal welfare in practice (Chapter 8). People are still free to believe that animals are consciously aware of what they are doing or, alternatively, not to believe it. But the definition itself is best left as having the two basic elements of health and what animals want and not to be burdened with the additional, mysterious, hard-to-define third element of consciousness.

The idea that although we know how to measure health, physiology and behaviour objectively, what we are 'really' after is what animals consciously feel has a detrimental effect on the whole of animal welfare science. It makes it look as though animal welfare scientists are forever chasing something they can never find and that all the measures they can come up with are nothing more than shadows or reflections of what they really mean. It leads to the erroneous conclusion that the solution is to take as many measures as possible and hope that the difficulties with individual measures will be somehow cancelled out by the presence of enough other, equally flawed, measures. It is, as argued in Chapter 2, one of the main reasons why it has proved so difficult to come up with the agreed definition of 'animal welfare' that is so urgently needed if the interests of non-human species are to count in a world that already has many competing human needs to balance.

Conclusions

A scientific approach to animal welfare needs a clear definition of what good welfare is, one that can be understood and agreed to by everyone, and, above all, one that points directly to the evidence that needs to be collected to evaluate welfare. 'Health and what animals want' provides such a definition. Trying to add a third element—what animals consciously feel—only makes the definition less clear, less useful and less likely to be universally adopted. This definition of animal welfare that makes no direct reference to conscious experiences still leaves open the possibility that many animals have vivid conscious experiences. But, by being grounded in what can actually be measured about their physiology and behaviour, it paves the way for genuine improvements in the lives of animals.

10

Consequences

It may seem odd to have another chapter after the Conclusions, but assessing animal welfare is only the first step on a long journey that each of us takes to decide our attitudes to non-human animals. Once we have established what keeps them healthy and what they want, we are faced with a range of practical problems about what do with that knowledge—for example, what attitude to take to farming, zoos, pest control, wildlife conservation, experiments and even to keeping pets in our homes. Whether we like it or not, we enter the worlds of politics, business and ethics. This book will not take you all the way on that journey. Its remit is strictly the scientific assessment of animal welfare, but this final chapter does contain some pointers to the way ahead—a map of the minefield, as it were—of how humans make up their minds on how to treat animals.

Animal welfare and ethics

Facts about animals, such as what keeps them healthy and what they want, do not in themselves tell you what you ought to do about those animals. One person might want to use a set of facts to make sure that animals have a good quality of life while they are alive, but also believe that it is acceptable to kill them for food as long as this is done humanely. Someone else might look at those same facts and conclude that it is quite unacceptable to kill them for food, however good their lives have been or however humanely they die. Same facts, different ethics.

In an ideal world, it would be possible to make a clear distinction between the science that tells you what is the case and the ethics that tells you what you ought to do, but in the real world this distinction becomes blurred. Science is never about pure, neutral facts because facts are discovered by people and people come with their own views and prejudices. Even the term 'animal welfare science' comes with its own built-in assumption that animal welfare is a desirable goal that we ought to be striving for. The science is carried out by a group of people who believe that animal welfare is important, in the same way that 'conservation science' is carried out by people who believe that we should be conserving natural habitats. These assumptions inevitably have an effect on the facts that are produced by the science, if only because they affect which questions get asked and therefore which answers become available. In addition, many people are convinced that the

The Science of Animal Welfare: Understanding What Animals Want. Marian Stamp Dawkins, Oxford University Press (2021).
 DOI: 10.1093/oso/9780198848981.003.0010

funding that a scientist receives affects the results they find so that they would get different results depending on whether their research was paid for by, say, an animal charity, the government or a drugs company. Certainly there does appear to be a problem with replicating the results of research in some areas of science (Baker 2016; Wyatt 2020), so no-one should see the findings of scientific research as fixed, immutable or entirely value-free. Nevertheless, it can be a valuable contribution to debates about animal welfare (as well as our own thinking) if we make an attempt to separate disagreements about the facts we learn from animal welfare science from disagreements about ethical values, and to separate both from the practical steps that might be taken in the real world to bring about change.

Animal welfare and animal rights

In discussions of animal ethics, a distinction is often made between animal welfare and animal rights. 'Animal welfare' refers to the view that what matters is how animals are treated while they are alive and that, as long as they have a good life, it is acceptable to kill them for food, to use them for experiments or for some other human purpose. 'Animal rights' refers to the belief that we should no more use animals in this way than we should use other human beings because animals have rights comparable to human rights, and one of those rights is the right not to be killed (Regan 1984). The more extreme versions of animal rights hold that animals and humans are equal, not master and slave, and that even domesticating animals is exploitation and should not happen (Francione 2004). Animals should be left alone in the wild to get on with their lives and neither need nor want humans to decide what is 'good' for them (Nussbaum 2004; Donaldson and Kymlicka 2011).

On the face of it, animal welfare and animal rights appear to be very different views, but, given the many different versions of each that have been expressed, it is probably more accurate to thinks of them as on a continuum with every possible shade of opinion in between (Beauchamp 2011). At one end of the spectrum are people who believe that humans and animals are morally of equal value and all have equal rights. But even Peter Singer, one of the leading proponents of animal rights, argued that 'There *are* important differences between humans and other animals that must give rise to *some* differences in the rights that each have' (Singer 1974, p. 569). So the belief that animals have rights still leaves the way open to a range of opinions about exactly what rights they have and which animals have them. On the welfarist side, we find a similar range of views from people who believe that humans have only a minimal duty to protect animals from unnecessary suffering (with a question mark over what is or is not 'unnecessary') to those who take their duty of protection to the point of being vegetarian or vegan.

To help you in your own thinking of where on the animal rights–animal welfare spectrum you are, you may find it useful to ask yourself what you would do if you were confronted with the philosophers' lifeboat dilemma. Philosophers like to use extreme examples to push us towards stating our moral position and then having to justify it. One of their favourites in animal ethics is a lifeboat on which there is limited remaining space.

The lifeboat is launched from a wrecked ship and there is just room in the lifeboat for either a large Labrador dog or one more human being, but not both. Which do you think should be allowed onto the lifeboat? The human? The dog? On what grounds? Is there a clear answer or should it be decided by a lottery in which the dog and the human are given an equal chance to survive?

Ethics and sentience

One concept that both unites and divides opinion is that of 'sentience', which has hung like a spectre over this whole book. Peter Singer (1976) argued that sentience—the capacity to experience suffering and enjoyment—should constitute the boundary of our ethical concern. If an organism is capable of suffering, he argued, its suffering is morally equivalent to the suffering of any other being. Taken literally, this would appear to say that the suffering of a human being is morally equivalent to that of a dog, a fish or even a crab, and that we ought therefore to give equal consideration to all of them. As mentioned above, however, even Singer does not take this extreme view and many people—both those coming from an animal welfare position and those holding animal rights views—have argued that, although important, sentience should not be the only criterion by which we decide which animals deserve our moral consideration (Frey 2011). As you might imagine, this has only spawned a further set of views about which other criteria should be used.

One view accepts that animals are sentient but says they are not 'persons' in the way that humans are and so do not deserve the same rights as humans (Raz 1984). 'Personhood' is then variously described as having the capacity for rational choice, abstract reasoning, language, long-term planning, and the ability to pursue desires and goals (Regan 1984; Donaldson and Kymlicka 2011; Tooley 2011). This allows some people to draw a distinction between humans and other animals but leaves them in the uncomfortable position that some humans such as babies or people with advanced dementia would not be counted as 'persons' either.

A crucial step in making ethical decisions about animals therefore becomes: what, if any, of the differences between humans and other animals are relevant to how they ought to be treated.

Speciesism

One word you may well come across in this context is 'speciesism', meaning discrimination against other animals because they do not belong to our species (Ryder 1975). Speciesism is the view that humans deserve special treatment over and above all other animals just because they belong to a particular species—namely our own. It has been compared to racism or sexism and criticized as equally unjustified (Singer 1976; Frey 2011). It is important to realize, however, that speciesism comes in several different guises. Some people are speciesist because, for religious or other reasons, they regard humans as superior to all other animals just by virtue of being human. They would see a human embryo, even one at a very early stage of development, as more valuable than an adult chimpanzee, however intelligent, because one is a member of the human species and the other is not.

More subtle, however, is speciesism when some reason is given for favouring humans, such as that animals lack some attribute that humans have and that, by implication, means that their suffering is ethically less important. One such reason might be that they are not 'persons', which, as indicated above, could include lacking the capacity to use language or to plan ahead for the future. But the problem with speciesism, even when it is linked to a particular attribute, is to justify the attribute as ethically relevant. What has language got to do with ethics? Is it that having language means that human suffer more because they can understand more what is happening to them? Or because having a language means they are more sentient? Or more intelligent? And does that mean that the more intelligent you are, the more moral value you have? And in any case, does being a 'person' make any difference at all to how you should be treated? This is not the place to give answers to such difficult questions. My purpose here is not to provide answers but simply to show you the complexity of the arguments involved and the importance of keeping facts about animal welfare (which we can hope to discover and even agree on) separate from the ethics of how we ought to treat animals (on which there are so many differing opinions).

One point is worth noting. Although there is an increasing number of countries around the world that now have laws giving protection to animals and valuing them as sentient beings, most legal systems are still speciesist in their attitudes to killing and death. Killing a human being 'to put them out of their misery', even if they were begging you to do so, would in many countries put you in jeopardy of being charged with murder. *Not* killing an animal that was suffering, on the other hand, would often result in prosecution. Many people regard suffering as worse than death for animals but death as worse than suffering for humans. Speciesism is thus deeply embedded in human attitudes to animals even where the long-term goal is to improve animal welfare.

Animal welfare and conservation ethics

It is not often realized that some of the most difficult ethical decisions about animal welfare occur in the context of wildlife conservation. Animal welfare is about the state of individuals, while conservation is about the health of whole ecosystems, and these two may come into conflict (Pacquet and Darimont 2010; Fraser and MacRae 2011). 'Biodiversity'—that much acclaimed goal of environmentalists—is often only achieved by ruthlessly killing off one species so that another species that is more preferred by humans can survive. Conservation, like animal welfare, is plagued by words that carry their own message about what ought to be done. Giving a species a label such as 'introduced', 'abundant' or 'pest' immediately implies that eliminating it can be justified and can obscure what that might mean for the welfare of the individual animals concerned

(Dubois et al. 2017). But of course a 'pest' animal can have the same pain, pleasure and suffering as a 'pet' or some iconic animal we humans happen to approve of.

This highlights a real problem that arises in putting animal ethics into practice. It may be relatively easy to convince people that good animal welfare is a desirable goal but much more difficult to resolve the conflicts that arise when what is good or desirable for one species clashes with what is good or desirable for others or for humans. If the welfare of humans depends on growing food that would otherwise be consumed by animals, whose welfare should take precedence? This conflict is of course not confined to conservation but arises wherever the interests of different parties come into conflict; for example, in using animals to test a vaccine that will save many human lives. Indeed, it is not even confined to human–animal conflicts but affects all areas of human life, such as when the welfare of one group of humans depends on their having access to land or resources that another group of humans believes 'ought' to be theirs. In other words, few ethical decisions about the welfare of one group can be taken without having implications for the welfare of other groups. To have maximum impact, therefore, animal ethics needs to be considered in the context of its implications for other values that society also holds. This means that 'ethical' comes to mean that all relevant concerns—including needs, benefits and costs to both people and animals—have been taken into account (Ramp and Bekoff 2015; Dubois et al. 2017). Of course this does not resolve the conflicts. There will still be arguments about whether the benefits of, say, preserving one species by killing its predators outweigh the costs to the predators of being killed, but at least it puts both species into the ethical picture and makes us consider whether there might be alternatives or ways in which the conflicts could be resolved to mutual benefit. Where the conflicts involve humans and other species, it turns out that there can be some surprising advantages to humans of valuing animal welfare.

Animal welfare and human welfare

For some people, the idea of making the case for animal welfare by appealing to what is good for humans is repellent and even unethical. It appears to devalue animals and to go against a deeply held belief that animals should be valued for themselves, not because they are useful to humans. But linking human and animal welfare need not imply any such devaluation. By looking for solutions in which what is good for animal welfare is also good for humans, animal welfare is given a powerful new set of arguments. Nothing is taken away. All that has happened is that if win–win situations can be found, everyone benefits, humans and non-humans too.

Conservationists long ago discovered the power of appealing to self-interest. Seeing ecosystems as 'natural capital' and stressing the economic and social value of clean water, healthy soil and unpolluted air, they can strengthen the case for taking care of the environment over and above valuing a forest for itself or for its intrinsic beauty (Balmford et al. 2002). It does not devalue a forest to say that it is beautiful *and* that it helps to improve the atmosphere. Similarly, it does not devalue animals to say that their welfare is important *and* that healthy, high-welfare animals also benefit humans, for example, by being less likely to harbour diseases than can infect humans (Queenan et al. 2016).

If there are human benefits to higher standards of animal welfare, this adds to, rather than detracts from, the arguments for improving animal welfare (Dawkins 2012; 2017) The more different arguments, including economic benefits (Norwood and Lusk 2011; Heise et al. 2018), that can be mustered in favour of better animal welfare, the better for the animals themselves. In this context, 'and' is a more powerful word than 'or'. The One Health programme (Monath et al. 2010; Ruegg et al. 2017) makes it clear that it is the combination of animal health and human health that matters and that animal health is a necessary condition for human health. There is also a growing view that good animal welfare is or should be an essential part of a sustainable way of life for humans (Garnett et al. 2013; Kelly et al. 2018). The more we can look for solutions in which humans and animals and the environment all gain, the more likely it is that higher standards of animal welfare will be achieved (Dawkins 2012; Llonch et al. 2017).

Defining animal welfare as what keeps animals healthy and what they want makes it particularly easy to link it to human benefits. Healthier animals mean healthier food for humans (Harley et al. 2012), less chance that animals develop a disease that will spread to humans, less wastage, less need for medication and higher-quality products. And, as we have seen throughout this book, there is a close relationship between whether animals have what they want and their health. There is increasing evidence that severely stressed animals—such as those deprived for long periods of what they want or forced to live in conditions that they do not want—develop the diseases of stress including reduced immune function (Ingvartsen and Moyes 2013). There are other less obvious benefits too. Attention to good welfare can bring better, more consistent experimental results, reducing the numbers of animals needed (Würbel 2009), farmers and zoo keepers get higher job satisfaction from looking after healthy contented animals and pets with high welfare will be better companions for people. Good animal welfare pays in all sorts of ways.

The road to improving animal welfare

Improving animal welfare involves much more than research findings, important though these are initially for providing the basic evidence. The research has then to be put into practice and this can often be a long and difficult process, requiring persistence and an ability to pivot in the face of difficulties and to adjust to circumstances (Grandin 2019). Scaling up from a well-designed, carefully controlled but small research project to commercial farming with all the uncertainties of weather, dust, humidity, power outages, large numbers of animals and other problems including human fallibility is not easy. The solutions must be financially viable because producers who attempt to adopt uneconomic solutions will go out of business. The public may have to be persuaded pay more for their food, and this may be easier for some than others, much more feasible for rich countries than for poorer ones. There are many other hazards along the way.

This book has been an attempt to make sure that the first step on that long road from identifying a welfare problem to seeing it solved in the real world is as sound and as sure-footed as possible. That first step is a clear and universally agreed definition of welfare, one that sets everything up for the journey ahead. 'Health and what animals want' specifies exactly what evidence we need to collect. It also provides the valence that is missing from so many 'measures' of welfare. At the same time, by being easy to understand, it also provides the public, farmers, politicians and everyone else with a way of seeing whether the evidence that is being collected is doing the job asked of it and so prepares the way for future implementation. If everyone can buy into the same meaning of what good welfare is, then there is a real chance of being able to agree on what should be done about it, how conflicts could be resolved and so of bringing about genuine improvements in the welfare of animals.

References

Abeyesinghe S M, Wathes C M, Nicol, C J and Randall J M (2001) The aversion of broiler chickens to concurrent vibrational and thermal stressors. *Applied Animal Behaviour Science* 73, 199–215.

Ahloy-Dallaire J, Espinosa J and Mason G (2017) Play and optimal welfare: does play indicate the presence of positive affective states? *Behavioural Processes* 156, 3–15.

Aldara-Kane A, Angulo F J, Conly J, Minato Y, Silbergeld E K, McEwen S A and Collignon P J (2018) World Health Organization (WHO) guidelines on use of medically important microbials in food-processing animals. *Antimicrobial Resistance and Infection Control* 7, article 7.

Anderson D J and Adolphs R (2014) A framework for studying emotions across species. *Cell* 157, 187–200.

Anderson A K and Phelps E A (2002) Is the human amygdala critical for subjective experience of emotion? Evidence of intact dispositional affect in patients with amygdala lesions. *Journal of Cognitive Neuroscience* 14(5), 709–20.

Andrew R J (1964) Vocalisation in chicks and the concept of 'stimulus contrast'. *Animal Behaviour* 12, 64–76.

Anselme P and Robinson M J F (2016) 'Wanting', 'liking', and their relation to consciousness. *The Journal of Experimental Psychology: Animal Learning and Cognition* 42, 123–40.

Arlinghaus R, Schweb A, Cooke S J and Cowx I G (2009) Contrasting pragmatic and suffering-centred approaches to fish welfare on recreational angling. *Journal of Fish Biology* 75, 2448–63.

Arranz L, de Vicente A, Muneoz M and de la Fuenta M (2009) Impaired immune function in a homeless population with stress-related disorders. *Neuroimmunomodulation* 16, 251–60.

Axelrod V, Bar M and Rees G (2015) Exploring the unconscious using faces. *Trends in Cognitive Sciences* 19, 35–45.

Baars B J (1988) *A Cognitive Theory of Consciousness*. Cambridge University Press, Cambridge.

Baars B J (2005) Subjective experience is probably not limited to humans: the evidence from neurobiology and behaviour. *Consciousness and Cognition* 14, 7–21.

Baars B J, Franklin S and Ramsoy T Z (2013) Global workspace dynamics: cortical 'binding and propagation' enables conscious contents. *Frontiers in Psychology* 4, article 200.

Baciadonna L and McElliot A G (2015) The use of judgement bias to assess welfare in farm livestock. *Animal Welfare* 24(1), 81–91.

Baker M (2016) Is there a reproducibility crisis? *Nature* 533, 452–4.

Balcombe J (2006) *Pleasurable Kingdom: Animals and the Nature of Feeling Good.* Macmillan, London.

Baldwin B A and Meese G B (1977) Sensory reinforcement and illumination preferences in the domesticated pig. *Animal Behaviour* 25(5), 497–507.

Baldwin B A and Start I B (1981) Sensory reinforcement and illumination preferences in sheep and calves. *Proceedings of the Royal Society* B 211, S13–26.

Balleine B W and Dickinson A (1998) Goal-directed instrumental action: contingency and incentive learning and their cortical substrates. *Neuropharmacology* 37, 407–19.

Balme G, Hunter L and Slotow R (2007) Feeding habitat selection by hunting leopards *Panthera pardus* in a woodland savanna: prey catchability versus abundance. *Animal Behaviour* 74, 589–98.

Balmford A, Bruner A, Cooper P, Costanza R, Farber S, Green R et al. (2002) Economic reasons for conserving wild nature. *Science* 297, 950–3.

Bardo, M T and Bevins R A (2000) Conditioned place preference: what does it add to our understanding of drug reward? *Psychopharmacology* 153, 31–43.

Barnett J L (1987) The physiological concept of stress is useful for assessing welfare. *Australian Veterinary Journal* 64, 195–6.

Barnett J L and Hemsworth P H (1990) The validity of physiological and behavioural measures of animal welfare. *Applied Animal Behaviour Science* 25, 177–87.

Barron A B and Klein C (2016) What insects can tell us about the origins of consciousness. *Proceedings of the National Academy of Sciences* 113, 4900–8.

Bateson M (2016) Optimistic and pessimistic biases: a primer for behavioural ecologists. *Current Opinion in Behavioral Sciences* 12, 115–21.

Bateson M and Matheson S M (2007) Performance on a categorization task suggests that removal of environmental enrichment induces 'pessimism' in captive European starlings (*Sturnus vulgaris*). *Animal Welfare* 16, 33–6.

Bateson P and Young M (1981) Separation from the mother and the development of play in cats. *Animal Behaviour* 29, 173–80.

Bauer E B and Smuts B B (2007) Cooperation and competition during dyadic play in domestic dogs, *Canis familaris. Animal Behaviour* 73, 489–99.

Baumans V and Van Loo P L P (2013) How to improve housing conditions of laboratory animals: the possibilities of environmental refinement. *Veterinary Journal* 195, 24–32.

Beauchamp T (2011) Rights theory and animal rights. In T Beauchamp and R Frey (eds) *The Oxford Handbook of Animal Ethics*. Oxford University Press, Oxford, pp. 198–227.

Beerling W, Koolhaas, J M, Ahnaou A, Bouwknecht J A, de Boer S F, Meerlo P et al. (2011) Physiological and hormonal responses to novelty exposure in rats are mainly related to ongoing behavioural activity. *Physiology and Behavior* 103, 412–20.

Bekoff, M. (2007) *The Emotional Lives of Animals*. New World Library, Novato, CA.

Bentham J (1789) (1961) Introduction to the principles of morals and legislation. In G Sher (ed.) *The Utilitarians* (2001). Hackett Publishing Co., Indianapolis, IN.

Berghman L R (2016) Immune responses to improving welfare. *Poultry Science* 95, 2216–18.

Bermond B (2001) A neuropsychological and evolutionary approach to animal consciousness and animal suffering. *Animal Welfare* 10, S47–62.

Berridge K C (2000) Measuring hedonic impact in animals and infants: microstructure of affective taste reactivity patterns. *Neuroscience and Biobehavioral Reviews* 24, 173–98.

Berridge K C and Robinson T E (1998) What is role of dopamine in reward: hedonic impact, reward learning, or incentive salience? *Brain Research Reviews* 28, 309–69.

Berridge K C, Robinson T E and Aldridge J W (2009) Dissecting components of reward: 'liking', 'wanting' and learning. *Current Opinion in Pharmacology* 9, 65–73.

Bethell E (2015) A 'how-to' guide for designing judgement bias studies to assess captive animal welfare. *Journal of Applied Animal Welfare Science* 18(Suppl 1), S18–42.

Birch J (2017) Animal sentience and the precautionary principle. *Animal Sentience* 2(16), article 017.

Blackmore S and Troscianko E T (2018) *Consciousness. An Introduction.* 3rd ed. Routledge, Abingdon.

Blanchard R J, Blanchard D C, Agullana R and Weiss S M (1990) 22kHz alarm cries in the laboratory rat. *Physiology and Behavior* 50, 967–72.

Block N (1995) On a confusion about the function of consciousness. *Behavioral and Brain Sciences* 18, 227–47.

Bloomfield R C, Gillespie G R, Kerswell K J, Butler K L and Hemsworth P H (2015) Effect of partial covering of the visitor viewing area window on positioning and orientation of zoo orangutans: a preference test. *Zoo Biology* 34, 223–29.

Boakes R (1984) *From Darwin to Behaviorism: Psychology and the Minds of Animals.* Cambridge University Press, Cambridge.

Boissy A and LeNeindre, P (1997) Behavioral, cardiac and cortisol responses to brief peer separation and reunion in cattle. *Physiology and Behavior* 61(5), 693–9.

Boissy A, Manteuffel G, Jensen M B, Moe R O, Spruijt B, Keeling L J et al. (2007) Assessment of positive emotions in animals to improve their welfare. *Physiology and Behavior* 92, 375–97.

Bokkers E A M, Zimmerman, P H, Rodenberg T B and Koene P (2007) Walking behaviour of heavy and light broilers in an operant runway test with varying durations of feed deprivation and feed access. *Applied Animal Behaviour Science* 108, 129–42.

Boly M, Seth A K, Wilke M, Ingmundsen P, Baars B, Laureys S, Edelman D and Tsuchiya N (2013) Consciousness in humans and non-human animals: recent advances and future directions. *Frontiers in Psychology* 4, article 625.

Bonilla-Jaime A, Vásquez-Palacios G, Arteaga-Silva M and Retana-Márquez S (2006) Hormonal responses to different sexually-related conditions in male rats. *Hormones and Behavior* 49(3), 376–82.

Bono G and Mori B D (2005) Animals and their quality of life: considerations beyond mere welfare. *Veterinary Research Communications* 29(Suppl 2), 165–8.

Bootsma M, Swenne C A, and Van Bolhuis (1994) Heart rate and heart rate variability as indexes of sympathovagal balance. *American Journal of Physiology* 266, H1565–71.

Borell E von, Langbein J, Després G, Hansen S, Leterrier C, Marchent-Forde J et al. (2007) Heart rate variability as a measure of autonomic regulation of cardiac activity for assessing stress and welfare in farm animals—a review. *Physiology & Behavior* 92, 293–316.

Borstel von U U, Duncan I J H, Shoveller A K, Merkies K, Keeling L J and Millman S T (2009) Impact of riding in a coercively Rollkur posture on welfare and fear on performance horses. *Applied Animal Behaviour Science* 116, 228–36.

Bracke M B M and Hopster H (2006) Assessing the importance of natural behaviour for animal welfare. *Journal of Agricultural and Environmental Ethics* 19, 77–89.

Bradshaw A L, and Poling A (1991) Choice by rats for enriched versus standard home cages: plastic pipes, wood platforms, wood chips, and paper towels as enrichment items. *Journal of the Experimental Analysis of Behavior* 55, 245–50.

Braithwaite V (2010) *Do Fish Feel Pain?* Oxford University Press, Oxford.

Brambell F W R (Chairman) (1965) *Report of the Technical Committee to Enquire into the Welfare of Animals kept under Intensive Livestock Husbandry Systems.* Cmnd. 2836. Her Majesty's Stationery Office, London.

Brando S and Buchanan-Smith H M (2018) The 24/7 approach to promoting optimal welfare for captive wild animals. *Behavioural Processes* 156, 83–95.

Bray G A (2004) How do we get fat? An epidemiological and metabolic approach. *Clinics in Dermatology* 22, 281–8.

Breuner C W and Berk S A (2019) Using the van Nordwijk and de Jong Resource framework to evaluate glucocorticoid-fitness hypothesis. *Integrative and Comparative Biology* 59, 243–50.

Bromberg-Martin E S, Matsumoto M and Hikosaka O (2010) Dopamine in motivational control: rewarding, aversive and alerting. *Neuron* 68, 815–34.
Bronfman Z Z and Ginsberg S (2016) The transition to minimal consciousness through the evolution of association learning. *Frontiers in Psychology* 7, article 1954.
Bronson P H and Desjardins C (1982) Endocrine responses to sexual arousal in male mice. *Endocrinology* 111, 1286–91.
Broom D M (1998) Welfare, stress and the evolution of feelings. *Advances in the Study of Behavior* 27, 371–403.
Broom D M (2007) Quality of life means welfare: how is it related to other concepts and assessed? *Animal Welfare* 16, 43–53.
Broom D M (2014) *Sentience and Animal Welfare.* CAB International, Wallingford.
Broom D M and Johnson K G (1993) *Stress and Animal Welfare.* Chapman and Hall, London.
Brown R, Lau H and LeDoux J (2019) Understanding the higher order approach to consciousness. *Trends in Cognitive Sciences* 23(9), 754–68.
Bubier, N E (1996) The behavioural priorities of laying hens: the effect of cost/no cost multi-choice tests on time budgets. *Behavioural Processes* 37, 225–38.
Buijs S, Keeling L J, Vangestel C, Baert J and Tuyttens F A M (2011a) Neighbourhood analysis as an indicator of spatial requirements of broiler chickens. *Applied Animal Behaviour Science* 129, 111–20.
Buijs S, Keeling L J, Vangestel C, Baert J, Vangeyte J and Tuyttens F A M (2011b) Assessing attraction or avoidance between rabbits: comparison of distance-based methods to analyse spatial distribution. *Animal Behaviour* 82, 1235–43.
Buijs S, Ampe B and Tuyttens F A M (2017) Sensitivity of the Welfare Quality® broiler chicken protocol to differences between intensively reared indoor flocks: which factors explain overall classification? *Animal* 11(2), 244–53.
Burgdorf J and Panksepp J (2001) Tickling induces reward in adolescent rats. *Physiology and Behavior* 72, 167–73.
Burghardt G (2005) *The Genesis of Animal Play: Testing the Limits.* MIT Press, Cambridge, MA.
Burman O H P, Parker R M A, Paul E S and Mendl M (2008) A spatial judgement task to determine background emotional state in laboratory rats, *Rattus norvegicus. Animal Behaviour* 76, 801–9.
Buwalda B, Scholte J, de Boer S F, Coppeus C M and Koolhaas J M (2012) The acute glucocorticoid response does not differentiate between rewarding and aversive social stimuli in rats. *Hormones and Behavior* 61(2), 218–26.
Cabanac M (1992) Pleasure: the common currency. *Journal of Theoretical Biology* 155(2), 173–200.
Cabanac M, Cabanac A J and Parent A (2009) The emergence of consciousness in phylogeny. *Behavioral Brain Research* 198, 267–72.
Cannon W B (1929) *Bodily Changes in Pain, Hunger, Fear and Rage.* 2nd ed. Appleton, New York.
Cardoso C S, von Keyserlingk M A G, Hötzel M J, Robbins J and Weary D M (2018) Hot and bothered: public attitudes towards heat stress and outdoor access for dairy cows. *PLoS One* 13(10), article e0205352.
Carruthers P (2019) *Human and Animal Minds.* Oxford University Press, Oxford.
CDC (Center for Disease Control and Prevention) (2012) NIOSH Pocket Guide to Chemical Hazards, 2012 ed. Available at: http://www.cdc.gov/niosh/npg/npgd0028.html (accessed 17 August 2020).
Chalmers D J (1995) Facing up to the problem of consciousness. *Journal of Consciousness Studies* 2, 200–19.

Chalmers D J (2016) Panpsychism and panprotopsychism. In Bruntrup G and Jaskolla L (eds) *Panpsychism: Contemporary Perspectives*. Oxford University Press, New York, pp. 19–47.

Charlton G L and Rutter SM (2017) The behaviour of housed dairy cattle with and without pasture access: A review. *Applied Animal Behaviour Science* 192, 2–9.

Charlton G L, Rutter S M, East M and Sinclair L A (2011a) Effects of providing total mixed rations indoors and on pasture on the behaviour of lactating dairy and their preferences to be indoors or on pasture. *Journal of Dairy Science* 94(8), 3875–84.

Charlton G L, Rutter S M, East M and Sinclair L A (2011b) Preference of dairy cows: indoor cubicle housing with access to a total mixed ration vs. access to pasture. *Applied Animal Behaviour Science* 130, 1–9.

Chen S and Sato S (2017) Role of oxytocin in improving the welfare of farm animals—a review. *Asian-Australian Journal of Animal Sciences* 30, 449–54.

Clapp J B, Croakin S, Dolphin C and Lyons S K (2015) Heart rate variability: a biomarker of dairy calf welfare. *Animal Production Science* 55(10), 1289–94.

Cleeremans A, Achoui D, Beauny A, Keumimka L, Martin J-R, Muñoz-Molden S et al. (2020) Learning to be conscious. *Trends in Cognitive Sciences* 24, 112–23.

Clubb R and Mason G (2003) Captivity effects on wide-ranging carnivores. *Nature* 425, 473–4.

Clubb R and Mason G J (2004) Pacing polar bears and stoical sheep: testing ecological and evolutionary hypotheses about animal welfare. *Animal Welfare* 13, 33–40.

Clubb R and Mason G J (2007) Natural behavioural biology as a risk factor in carnivore welfare: how analysing species differences could help zoos improve enclosures. *Applied Animal Behaviour Science* 102, 303–28.

Cockram M S (2004) A review of behavioural and physiological responses of sheep to stressors to identify potential behavioural signs of distress. *Animal Welfare* 13(3), 283–91.

Cockram M S and Hughes B O (2011) Health and disease. In M C Appleby, J A Mench, I A S Olsson and B O Hughes (eds) *Animal Welfare*. 2nd ed. CABI, Wallingford, pp. 120–37.

Colditz I G and Hine B C (2016) Resilience in farm animals: biology, management, breeding and implications for animal welfare. *Animal Production Science* 56, 1961–83.

Collias N E and Joos M (1953) The spectrographic analysis of sound signals of the domestic fowl. *Behaviour* 5, 175–88.

Collins T, Stockman C A, Barnes A L, Miller D W, Wickham S L and Fleming P A (2018) Qualitative behavioural assessment as a method to identify potential stressors during commercial sheep transport. *Animals* 8, article 209.

Colman A M, Fulford B D, Omtzigt D and al-Nowaihi A (2010) Learning to cooperate without awareness in multiplayer minimal social situations. *Cognitive Psychology* 61, 201–27.

Courcier E A, Thomson R M, Mellor D J and Yam P S (2010) An epidemiological study of environmental factors associated with canine obesity. *Journal of Small Animal Practice* 51, 362–7.

Cronin K A and Ross S R (2019) A cautionary note on the use of behavioural diversity (H-index) in animal welfare science. *Animal Welfare* 28, 157–64.

Cyr N E, Earle K, Tam C and Romero M (2007) The effect of chronic psychological stress on corticosterone, plasma metabolites and immune responses in European starlings. *General and Comparative Endocrinology* 154, 59–66.

Dantzer R and Mormède P (1983) Stress in farm animals: a need for re-evaluation. *Journal of Animal Science* 57, 6–18.

Darwin C (1856) Letter to J.D. Hooker dated 13 July 1856. Letter no. DCP-LETT-1924, Darwin Correspondence Project, University of Cambridge. Available at https://www.darwinproject.ac.uk/ (accessed 28 April 2020).

Darwin C (1872) *The Expression of the Emotions in Man and Animals*. Reprinted 1965. University of Chicago Press Ltd, London.
Dawkins M S (1980) *Animal Suffering: The Science of Animal Welfare*. Chapman and Hall, London.
Dawkins M S (1983) Battery hens name their price: consumer demand theory and the measurement of ethological 'needs'. *Animal Behaviour* 31, 1195–1205.
Dawkins M S (1990) From an animal's point of view: motivation, fitness and animal welfare. *Behavior and Brain Sciences* 13, 1–61.
Dawkins M S (1993) *Through Our Eyes Only? The Search for Animal Consciousness*. W. H. Freeman/ Spektrum, New York and Heidelberg.
Dawkins M S (2008) The science of animal suffering. *Ethology* 114, 937–45.
Dawkins M S (2012) *Why Animals Matter: Animal Consciousness, Animal Welfare and Human Well-being*. Oxford University Press, Oxford.
Dawkins M S (2015) Animal welfare and the paradox of animal consciousness. *Advances in the Study of Behavior* 47, 5–38.
Dawkins M S (2017) Animal welfare and efficient farming: is conflict inevitable? *Animal Production Science* 57, 201–6.
Dawkins M S (2017) Animal welfare with and without consciousness. *Journal of Zoology* 301, 1–10.
Dawkins M S (2019) Animal welfare as preventative medicine. *Animal Welfare* 26, 137–41.
Dawkins M S, Cook P A, Whittingham M J, Mansell K A and Harper A E (2003) What makes free-range broilers range? *In situ* measurements of habitat preference. *Animal Behaviour* 66(1), 151–60.
Dawkins M S, Donnelly C A and Jones T A (2004) Chicken welfare is influenced more by housing conditions than stocking density. *Nature* 427, 342–4.
Dawkins M S, Edmond A, Lord A and Solomon S (2004) Time course of changes in egg shell quality, faecal corticosterones and behaviour as welfare measures in laying hens. *Animal Welfare* 13, 321–7.
Dawkins R (1982) *The Extended Phenotype*. Chapter 2. W. H. Freeman, Oxford.
Dawkins R and Dawkins M (1973) Decisions and the uncertainty of behaviour. *Behaviour* 45, 83–103.
Day J E L, Kyriazakis I, and Rogers P J (1997) Feeding motivation in animals and humans: a comparative review of its measurements and uses. *Nutrition Abstracts and Reviews* B 67, 69–79.
D'Eath R B, Tolkamp B J, Kyriazakis I and Lawrence A B (2009) 'Freedom from hunger' and preventing obesity: the animal welfare implications of reducing food quantity or quality. *Animal Behaviour* 77, 275–88.
Dehaene S (2014) *Consciousness and the Brain: Deciphering How the Brain Codes Our Thoughts*. Penguin Books, New York.
Dehaene S and Changeux J P (2011) Experimental and theoretical approaches to conscious processing. *Neuron* 70, 200–27.
Dehaene S, Naccache L, Cohen L, Le Bihan D, Mangin J-F, Poline J-P et al. (2001) Cerebral mechanisms of word masking in unconscious repetition priming. *Nature Neuroscience* 4, 752–8.
Dennett D (1991) *Consciousness Explained*. Little, Brown and Co., London.
Denton D A, McKinley M J, Farrell M and Egan G F (2009) The role of primordial emotions in the evolution of consciousness. *Consciousness and Cognition* 18, 500–14.
Devinney B J, Berman C M and Rasmussen K L R (2003) Individual differences in response to sibling birth among free-ranging yearling rhesus monkeys (*Macaca mulatta*) on Cayo Santiago. *Behaviour* 140, 899–924.
Dickinson A (2012) Associative learning and animal cognition. *Philosophical Transactions of the Royal Society* B 367, 2733–42.
Dimberg U, Thunberg M and Elmehed K (2000) Unconscious facial reactions to emotional facial expressions. *Psychological Science* 11, 86–9.

Dixon L M, Sandilands V, Bateson M, Brocklehurst S, Tolkamp B J and D'Eath R B (2013) Conditioned place preference or aversion as animal welfare assessment tools: limitations in their application. *Applied Animal Behaviour Science* 148, 164–76.

Dixon L M, Brocklehurst S, Sandilands V, Bateson M, Tolkamp, B J and D'Eath R B (2014) Measuring motivation for appetitive behaviour: food-restricted broiler breeder chickens cross a water barrier to forage in an area of wood shavings without food. *PLoS One* 9, article e102322.

Donaldson S and Kymlicka (2011) *Zoopolis. A Political Theory of Animal Rights*. Oxford University Press, Oxford.

Dubois S, Fenwick N, Ryan E A, Baker L, Baker S E, Beausoleiel N J et al. (2017) International consensus principles for ethical wildlife control. *Conservation Practice and Policy* 31, 753–60.

Duncan I J H (1992) Measuring preferences and strength of preferences. *Poultry Science* 71, 658–63.

Duncan, I J H (1993) Welfare is to do with what animals feel. *Journal of Agricultural and Environmental Ethics* 6(2), 8–14.

Ede T, Lecorps B, von Keyserlingk M A G and Weary D M (2019) Scientific assessment of affective states in dairy cattle. *Journal of Dairy Science* 102(11), 10677–94.

Edgar J L, Nicol C J, Pugh C A and Paul E S (2013) Surface temperature changes in response to handling in domestic chickens. *Physiology and Behavior* 119, 195–200.

European Union (2009) Lisbon Treaty. Available at: https://ec.europa.eu/food/animals/welfare_en (accessed 17 August 2020).

Evers K and Sigman M (2013) Possibilities and limits of mind-reading: a neurophilosophical persepective. *Consciousness and Cognition* 22, 887–97.

Fagen R (1981) *Animal Play Behavior*. Oxford University Press, Oxford.

Falk A C, Wear D M, Winckler C and Keyserlingk M A G (2012) Preference for pasture versus freestall housing by dairy cattle when stall availability indoors is reduced. *Journal of Dairy Science* 95, 6409–15.

Faure M and Lagadic H (1994) Elasticity of demand for food and sand in laying hens subjected to variable wind speed. *Applied Animal Behaviour Science* 42, 49–59.

FAWC (Farm Animal Welfare Council) (1979) Press statement, 5 December 1979. FAWC, London.

FAWC (Farm Animal Welfare Council) (2009) *Farm Animal Welfare in Great Britain: Past, Present and future*. FAWC, London.

Febrer K, Jones T A, Donnelly C A and Dawkins M S (2006) Forced to crowd or choosing to cluster? Spatial distribution indicates social attraction in broiler chickens. *Animal Behaviour* 72, 1291–300.

Feinberg T E and Mallatt J (2019) Subjectivity 'demystified': neurobiology, evolution and the explanatory gap. *Frontiers in Psychology* 10, article 1686.

Fischer D B, Boes A D, Demertzi A, Evrard H C, Laureys S and Edlaw B L (2016) A human brain network derived from coma-causing brain stem lesions. *Neurology* 87(23), 2427–34.

Fiorillo C D, Tobler P N and Schultz W (2003) Discrete coding of reward probability and uncertainty by dopamine neurons. *Science* 299, 1898–902.

Francione G L (2004) Animals—property or persons? In C R Sunstein and M C Nussbaum (eds) *Animal Rights*. Oxford University Press, Oxford, pp. 108–42.

Franks B (2019) What animals want. *Animal Welfare* 28, 1–10.

Fraser A F and Broom D M (1990) *Farm Animal Behaviour and Welfare*. 3rd ed. Baillière Tindall, London.

Fraser D (2008) *Understanding Animal Welfare: The Science in its Cultural Context*. Universities Federation for Animal Welfare, Wiley-Blackwell, Chichester.

Fraser D and MacRae A M (2011) Four types of activities that affect animals: implications for animal welfare science and animal ethics philosophy. *Animal Welfare* 20, 581–90.
Fraser D and Matthews L R (1997) Preference and motivation testing. In M C Appleby and B O Hughes (eds) *Animal Welfare*. CAB International, Wallingford, pp. 159–73.
Fraser D and Nicol C J (2011) Preference and motivation research. In M C Appleby, J A Mench, I A S Olsson and B O Hughes (eds) *Animal Welfare*. 2nd ed. CAB International, Wallingford, pp. 183–99.
Fraser D, Duncan I J H, Edwards S A, Grandin T, Gregory, NG et al. (2013) General principles for the welfare of animals in production systems: the underlying science and its application. *The Veterinary Journal* 198, 19–27.
Frey R G (2011) Utilitarianism and animals. In T L Beauchamp and R G Frey (eds) *The Oxford Handbook of Animal Ethics*. Oxford University Press, Oxford, pp. 172–97.
Friel M, Kunc H P, Griffin K, Asher L and Collins L M (2019) Positive and negative contexts predict duration of pig vocalisations. *Scientific Reports* 9, article 2062.
Friend T H, Dellmeier G R and Gbur E E (1985) Comparison of four methods of calf confinement. I. Physiology. *Journal of Animal Science* 60, 1095–101.
Garnett T, Appleby M C, Balmford A, Bateman J, Benton T, Bloomer P et al. (2013) Sustainable intensification in agriculture: premises and promises. *Science* 341, 32–4.
German A J (2006) The growing problem of obesity in dogs and cats. *Journal of Nutrition* 136, 1940S–6S.
Godfray H C J, Beddington J R, Crute I R, Haddad L, Lawrence D, Muir J F et al. (2010) Food security: the challenge of feeding 9 billion people. *Science* 327, 812–17.
Gomez Y, Bieler R, Hankele A K, Zähner M, Savary P and Hillmann E (2018) Evaluation of visible eye white and maximum eye temperature as non-invasive indicators of stress in dairy cows. *Applied Animal Behaviour Science* 198, 1–8.
Goodall J (1990) *Through a Window*. Houghton-Mifflin, Boston, MA.
Grandin T (2019) Crossing the divide between academic research and practical application of ethology and animal behaviour information in commercial livestock and poultry farms. *Applied Animal Behaviour Science* 218, article UNS 104828.
Greco B J, Meehan C L, Heinsius J L and Mench J A (2017) Why pace? The influence of social, housing, management, life history, and demographic characteristics on locomotor stereotypy in zoo elephants. *Applied Animal Behaviour Science* 194, 104–11.
Green T C and Mellor D J (2011) Extending ideas about animal welfare assessment to include 'quality of life' and related concepts. *New Zealand Veterinary Journal* 59(6), 263–71.
Griffin D (1976) *The Question of Animal Awareness: Evolutionary Continuity of Animal Awareness*. Rockefeller University Press, New York.
Grill-Spector K T, Kushnir T, Nendler T and Malach R (2000) The dynamics of object-selective activation correlates with recognition performance in humans. *Nature Neuroscience* 3(8), 837–43.
Gross W B and Siegel P B (1981) Long-term exposure of chickens to three levels of social stress. *Avian Disease* 25, 312–26.
Guedes S R, Valentim A M and Antunes L M (2017) Mice aversion to sevoflurane, isoflurane and carbon dioxide using an approach-avoidance task. *Applied Animal Behaviour Science* 189, 91–7.
Gunnarsson S, Keeling L J and Svedberg J (1999) Effect of rearing factors on the prevalence of feather-pecking in commercial flocks of loose housed laying hens. *British Poultry Science* 40, 12–18.
Gunnarsson S, Matthews L R, Foster T M and Temple W (2000) The demand for straw and feathers as litter substrates by laying hens. *Applied Animal Behaviour Science* 65, 321–30.

Guyenet S J and Schwartz M W (2012) Clinical review: Regulation of food intake, energy balance, and body fat mass: implication for the pathogenesis and treatment of obesity. *Journal of Clinical Endrocrinology and Metabolism* 97, 745–55.

Gutfreund Y (2018) The mind-evolution problem: the difficulty of fitting consciousness in an evolutionary framework. *Frontiers in Psychology* 9, article 1537.

Gygax L (2017) Wanting, liking and welfare: the role of affective states in proximate control of behaviour in vertebrates. *Ethology* 123, 689–704.

Gygax L and Hillman E (2018) 'Naturalness' and its relation to animal welfare from an ethological perspective. *Agriculture* 8, 136.

Hanmer L A, Riddell P M and Williams C M (2010) Using a runway paradigm to assess the relative strength of rats' motivations for enrichment objects. *Behavior Research Methods* 42, 517–24.

Hansen S and Jeppesen L L (2006) Temperament, stereotypies and anticipatory behaviour as measures of welfare in mink. *Applied Animal Behaviour Science* 99, 172–82.

Harding E J, Paul E S and Mendl M (2004) Animal behaviour—cognitive bias and affective state. *Nature* 427, 312–12.

Harley S, More S, Boyle L, O'Connell N and Hanlon A (2012) Good animal welfare makes economic sense: potential of pig abattoir meat inspection as a welfare surveillance tool. *Irish Veterinary Journal* 65, article 11.

Haskell M, Wemelsfelder F, Mendl M, Calvert S and Lawrence A B (1996) The effect of substrate-enriched and substrate-impoverished housing environments in the diversity of behaviour in pigs. *Behaviour* 133, 741–61.

Haskell S P, Ballard W B, Mcroberts J T, Wallace M C, Krausman P R, Humphrey M H et al. (2017) Growth and mortality of sympatric white-tailed and mule deer fawns. *Journal of Wildlife Management* 81(8), 1417–29.

Hayes J E, McGreevy P D, Forbes S L, Laing G and Stuetz R M (2018) Critical review of dog detection and the influences of physiology, training, and analytical methodologies. *Talanta* 185, 499–512.

Held S D E and Špinka M (2011) Animal play and animal welfare. *Animal Behaviour* 81(5), 891–9.

Heimeberge S, Kanitz E and Otten W (2019) The use of hair cortisol for the assessment of stress in animals. *General and Comparative Endocrinology* 270, 10–17.

Heise H, Schwarze S and Theuvsen L (2018) Economic effects of participation in animal welfare programmes: does it pay off for farmers? *Animal Welfare* 27, 167–79.

Henry J P (1977) *Stress, Health and the Social Environment*. Springer-Verlag, New York.

Herborn K A, Graves J L, Jerem P, Evans N P, Nager R, McCafferty D J et al. (2015) Skin temperature reveals the intensity of acute stress. *Physiology and Behavior* 152, 225–30.

Hernandez M C, Navarro-Castilla A, Planillo A, Sanchez-Gonzalez B and Barja I (2018) The landscape of fear: why some free-ranging rodents choose repeated live-trapping over predation risk and how it is associated with the physiological stress response *Behavioural Processes* 157, 125–32.

Hilborn A, Pettorelli N, Orme C D L and Durant S M (2012) Stalk and chase: how hunt stages affect hunting success in Serengeti cheetah. *Animal Behaviour* 84, 701–6.

Hill J E, DeVault T L and Belant J L (2019) Cause-specific mortality of the world's terrestrial vertebrates. *Global Ecology and Biogeography* 28, 680–9.

Hill J O and Peters J C (1998) Environmental contributions to the obesity epidemic. *Science* 280, 1371–4.

Hintze S, Murphy E, Bachmann I, Wemelsfelder F and Wurbel H (2017) Qualitative Behavioural Assessment of horses exposed to short-term emotional treatments. *Applied Animal Behaviour Science* 196, 44–51.

Holm L, Jensen M B and Jeppesen L L (2002) Calves' motivation for access to two different types of social contact measured by operant conditioning. *Applied Animal Behaviour Science* 79, 175–94.

Horta O (2018) Concern for wild animal suffering and environmental ethics: what are the limits of the disagreement? *The Ethics Forum* 13, 85–100.

Houston A I, McNamara J M, Barta Z and Klasing K C (2007) The effect of energy reserves and food availability on optimal immune defence. *Proceedings of the Royal Society* B 274, 2835–47.

Houston A I, Trimmer P C, Fawcett T W, Higginson A D, Marshall J A R and McNamara J M (2012) Is optimism optimal? Functional causes of apparent behavioural biases. *Behavioural Processes* 89(7), 172–8.

Howell C P and Cheyne S M (2019) Complexities of using wild versus captive activity budget comparisons for assessing captive primate welfare. *Journal of Applied Animal Welfare Science* 22, 78–96.

Hudson J A, Frewer L J, Jones G, Brereton P A, Whittingham M J and Stewart G (2017) The agri-food chain and antimicrobial resistance: a review. *Trends in Food Science and Technology* 69, 131–47.

Hughes B O and Black A J (1973) The preference of domestic hens for different types of battery cage floor. *British Poultry Science* 14, 615–19.

Hughes B O, Gilbert A B and Brown M F (1986) Categorization and causes of abnormal egg-shells—relationship with stress. *British Poultry Science* 27, 325–37.

Hurnik J F (1993) Ethics and animal agriculture. *Journal of Agricultural and Environmental Ethics* 6(Suppl. 10), 21–35.

Ide J S, Nedic S, Wong K S, Story S L, Lawson E A et al. (2018) Oxytocin attenuates trust as a subset or more general reinforcement learning, with altered reward circuit connectivity in males. *Neuroimage* 174, 35–43.

Illius A W, Tolkamp B J and Yearsley J (2002) The evolution of the control of food intake. *Proceedings of the Nutrition Society* 61, 465–72.

Imfeld-Mueller S and Hillmann E (2012) Anticipation of a food ball increases short-term activity levels in growing pigs. *Applied Animal Behaviour Science* 137, 23–9.

Ingvartsen K L and Moyes K (2013) Nutrition, immune function and health of dairy cattle. *Animal* 7, 112–22.

Insabato A, Pannunzi M, Rolls E T and Deco G (2010) Confidence-related decision-making. *Journal of Neurophysiology* 104(1), 539–47.

Jensen M B (1999) Effects of confinement on rebounds of locomotory behaviour of calves and heifers, and the spatial preferences of calves. *Applied Animal Behaviour Science* 62, 43–56.

Jensen M B, Munksgaard L, Pedersen L J, Ladewig J and Matthews L (2004) Prior deprivation and reward duration affect the demand function for rest in dairy heifers. *Applied Animal Behaviour Science* 88, 1–11.

Jones E K M, Wathes C A and Webster A J F (2005) Avoidance of atmospheric ammonia by domestic fowl and the effect of early experience. *Applied Animal Behaviour Science* 90, 293–308.

Jones J B, Burgess L R Webster A J F and Wathes C M (1996) Behavioural responses of pigs to atmospheric ammonia in a chronic choice test. *Animal Science* 63, 437–45.

Jones R and Nicol C J (1998) A note on the effect of the thermal environment on the well-being of growing pigs. *Applied Animal Behaviour Science* 60(1), 1–9.

Kastrup B (2018) The universe of consciousness. *Journal of Consciousness Studies* 25, 125–55.

Katayama M, Kubo T, Mogi K, Ikeda K, Nagasawa M and Kikusui T (2016) Heart rate variability predicts the emotional state in dogs. *Behavioural Processes* 128, 108–12.

Kawamura G, Naohara T, Tanaka Y, Nishi T and Anraku K (2009) Near-ultraviolet radiation guides the emerged hatchlings of loggerhead turtles *Caretta caretta* (Linnaeus) from a nesting beach to the sea at night. *Marine and Freshwater Behaviour and Physiology* 42, 19–30.

Keeling L (1995) Spacing behaviour and an ethological approach to assessing optimal space allowances for laying hens. *Applied Animal Behaviour Science* 44, 171–86.

Keeling L J and Duncan I J H (1989) Inter-individual distances and orientation in laying hens housed in groups of 3 in 2 different sized enclosures. *Applied Animal Behaviour Science* 24(4), 325–42.

Keeling L J and Duncan, I J H (1991) Social spacing in domestic fowl under semi-natural conditions. The effect of behavioural activity and activity transitions. *Applied Animal Behaviour Science* 32, 205–17.

Keller M and Mustin W (2017) Examining the preference for shade structures in farmed green sea turtles (*Chelonia mydas*) and shade's effect on growth and temperatures. *Journal of Zoo and Wildlife Medicine* 48, 121–30.

Kelly E, Latruffe L, Desjeux Y, Ryan M, Uthes S, Diazabakana A et al. (2018) Sustainability indicators for improved assessment of agricultural policy across the EU: is FADN the answer? *Ecological Indicators* 89, 903–11.

Keltner D, Oatley K and Jenkins J M (2013) *Understanding Emotions*. Wiley-Blackwell, Oxford.

Kent J E and Ewbank R (1983) The effect of road transportation on the blood constituents and behaviour of calves.1. Six months old. *British Veterinary Journal* 139, 228–35.

Kilgour R, Foster T M, Temple W, Matthews L R and Bremner K J (1991) Operant technology applied to solving farm animal problems. An assessment. *Applied Animal Behaviour Science* 30, 141–66.

Kirkden R D, Edwards J S S and Broom D M (2003) A theoretical comparison of the consumer surplus and the elasticities of demand as measures of motivational strength. *Animal Behaviour* 65, 157–78.

Kirkden R D, Lee L, Lee G, Makowska I J, Pfaffinger M J and Weary D M (2008) The validity of using am approach-avoidance test to measure the strength of aversion to carbob dioxide in rats. *Applied Animal Behaviour Science* 114, 216–34.

Knutson B, Burgdorf J and Panksepp J (1999) High frequency ultrasonic vocalisations index conditioned pharmacological reward in rats. *Physiology and Behavior* 66, 639–43.

Koch, C (2004) *The Quest for Consciousness. A neurobiological approach*. Roberts and Co., Englewood, CO.

Koch C (2017) How to make a consciousness meter. *Scientific American* 317(5), 28–33.

Koch C, Massimini M, Boly M and Tononi G (2016) Neural correlates of consciousness: progress and problems. *Nature Reviews Neuroscience* 17, 307–21.

Koistinen T, Ahola L and Monenen J (2007) Blue foxes' motivation for access to an earth floor measured by operant conditioning. *Applied Animal Behaviour Science* 107, 328.

Koolhaas J M, Meerlo P, de Boer S F, Strubbe J H and Bohus B (1997) The temporal dynamics of the stress response. *Neuroscience and Biobehavioral Reviews* 21, 775–82.

Koolhaas J M, Bartomolomucci A, Buwalda B, de Boer S F, Flugge G, Korte S M et al. (2011) Stress revisited: a critical evaluation of the stress concept. *Neuroscience and Biobehavioral Reviews* 35, 1291–301.

Kopelman P G (2000) Obesity as a medical problem. *Nature* 404, 635–43.

Koren L, Whiteside A, Fahlman K, Ruckstuhl K, Kutz S, Checkley S et al. (2012) Cortisol and corticosterone independence in cortisol-dominant wildlife. *General Comparative Endocrinology* 177, 113–19.

Kovács L, Kézér F L, Jurkovich V, Kulcsár-Huszenicza M and Tözsér J (2015) Heart rate variability as an indicator of chronic stress caused by lameness in dairy cows. *PLoS One* 10, article e1034792.

Krachun C, Rushen J and de Pasille A M (2010) Play behaviour in dairy calves is reduced by weaning and by low energy intake. *Applied Animal Behaviour Science* 122, 71–6.

Krause J and Ruxton G D (2002) *Living in Groups*. Oxford University Press, Oxford.

Kristensen H H, Burgess L R, Demmers T G H and Wathes C M (2000) The preferences of laying hens for different concentrations of atmospheric ammonia. *Applied Animal Behaviour Science* 68, 307–18.

Kroshko J, Clubb R, Harper L, Mellor E, Moehrenschlager A and Mason G J (2016) Stereotypic route tracing in captive Carnivora is predicted by species-typical home range sizes and hunting styles. *Animal Behaviour* 117, 197–209.

Kruuk H (1964) Predators and anti-predator behaviour of the black-headed gull (*Larus ridibundus* L). *Behaviour Supplement* 11.

Kruuk H (1972) *The Spotted Hyena*. University of Chicago Press, Chicago, IL.

Kulli J and Koch C (1991) Does anaesthesia cause loss of consciousness? *Trends in Neurosciences* 14, 6–10.

Lambert H S and Carter G (2017) Looking into the eyes of a cow. Can eye whites be used as a measure of emotional state? *Applied Animal Behaviour Science* 186, 1–6.

Langbein J, Siebert K and Nuernberg G (2009) On the use of an automated learning device by group-housed dwarf goats: do goats seek cognitive challenges? *Applied Animal Behaviour Science* 120, 150–8.

Langford D, Bailey A, Chanda M L, Clarke S E, Drummond T E, Echols S et al. (2010) Coding of facial expressions of pain in the laboratory mouse. *Nature Methods* 7, 447–9.

Larsen H, Cronin G, Smith C L, Hemsworth P and Rault, J L (2017) Behaviour of free-range laying hens in distinct outdoor environments. *Animal Welfare* 26, 255–64.

Lawrence A (1987) Consumer demand theory and the assessment of animal welfare. *Animal Behaviour* 35, 293–5.

Lawrence A B and Terlouw E M C (1993) A review of the behavioral factors involved in the development and continued performance of stereotypic behaviours in pigs. *Journal of Animal Science* 71, 2815–25.

Lay D C, Friend T H, Bowers C L, Grissom K K and Jenkins O C (1992) A comparative physiological and behavioural study of freeze and hot-iron branding using dairy cows. *Journal of Animal Science* 70, 1121–5.

Learmonth M J (2019) Dilemmas for natural living concepts in zoo animal welfare. *Animals* 8(6), article 318.

LeDoux J E (2014) Coming to terms with fear. *Proceedings of the National Academy of Sciences* 111, 2871–6.

LeDoux J E and Hoffman S G (2018) The subjective experience of emotions: a fearful view. *Current Opinion in Behavioral Sciences* 19, 67–72.

LeDoux J E and Pine D S (2016) Using neuroscience to help understand fear and anxiety: a two-system framework. *American Journal of Psychiatry* 173, 1083–93.

Lee C, Fisher A D, Colditz I G, Lea J M and Ferguson D M (2013) Preference of beef cattle for feedlot or pasture environments. *Applied Animal Behaviour Science* 145, 53–9.

Leliveld L M C, Dupjan S, Tuchscherer A and Puppe B (2016) Behavioural and physiological measures indicate subtle variations in the emotional valence of young pigs. *Physiology and Behaviour* 157, 116–24.

Lieberman L S (2006) Evolutionary and anthropological perspectives on optimal foraging in obesogenic environments. *Appetite* 47, 3–9.

Limpus C and Kamrowski R L (2013) Ocean-finding in marine turtles: the importance of low horizon elevation as an orientation cue. *Behaviour* 150, 863–93.

Llonch P, Haskell M J, Dewhurst R J and Turner S P (2017) Current available strategies to mitigate greenhouse gas emissions in livestock systems: an animal welfare perspective. *Animal* 11, 274–84.

Loy J (1970) Behavioral responses of free-ranging rhesus monkeys to food shortage. *American Journal of Physiological Anthropology* 33, 253–71.

Macphail, E M (1987) The comparative psychology of intelligence. *Behavioral and Brain Sciences* 10, 645–95.

MacLeod K J, Krebs C J, Boonstra R and Sheriff M J (2018) Fear and lethality in snowshoe hares: the deadly effects of non-consumptive predation risk. *Oikos* 127(3), 375–80.

Magana M, Alonso J C, Martin C A, Bautista L M and Martin B (2010) Nest-site selection by great bustards (*Otis tarda*) suggests a trade-off between concealment and visibility. *Ibis* 152, 77–89.

Makowska J I and Weary D M (2013) Assessing the emotions of laboratory rats. *Applied Animal Behaviour Science* 148, 1–12.

Makowski J I, Niel L, Kirkden R D and Weary D M (2008) Rats show aversion to argon-induced hypoxia. *Applied Animal Behaviour Science* 114, 148–51.

Manser M B (2013) Semantic communication in vervet monkeys and other animals. *Animal Behaviour* 86, 491–6.

Manteuffel G, Puppe B and Schön P C (2004) Vocalization of farm animals as a measure of welfare. *Applied Animal Behaviour Science* 88, 163–82.

Margulis L (2001) The conscious cell. *Annals New York Academy of Science* 929, 55–70.

Mashour G A and Alkire M T (2013) Evolution of consciousness: phylogeny, ontogeny and emergence from general anaesthesia. *Proceedings of the National Academy of Sciences* 110, 10357–64.

Mason G J (1991) Stereotypies: a critical review. *Animal Behaviour* 41, 1015–37.

Mason G J (2010) Species differences in responses to captivity: stress, welfare and the comparative method. *Trends in Ecology and Evolution* 25, 713–25.

Mason G J, Cooper J and Clarebrough C (2001) Frustrations of fur-farmed mink. *Nature* 410, 35–6.

Mason G J and Latham N R (2004) Can't stop, won't stop: is stereotypy a reliable animal welfare indicator? *Animal Welfare* 13, S57–69.

Mason G J and Mendl M (1993) Why is there no simple way of measuring animal welfare. *Animal Welfare* 2(4), 301–19.

Mason G J, McFarland D J and Garner J (1998) A demanding task: using economic techniques to assess animal priorities. *Animal Behaviour* 55, 1071–5.

Mason W A (1974) Specificity in the organization of neuroendocrine response profiles. In P Seemand and G M Brown (eds) *Frontiers in Neurology and Neuroscience Research*. University of Toronto Press, Toronto, pp. 68–80.

Mason W A, Sharpe L G and Saxon S V (1963) Preferential responses of young chimpanzees to food and social rewards. *Psychological Record* 13, 31–345.

Matthews L R and Ladewig J (1994) Environmental requirements of pigs measured by behavioural demand functions. *Animal Behaviour* 47, 713–19.

McBride G, James J W and Shaffner R N (1963) Social forces determining spacing and head orientation in a flock of domestic hens. *Nature* 197, 1272–3.

McConnachie E, Smid A M C, Thompson A J, Weary D M, Gaworski M A and von Keyserlingk, M A G (2018) Cows are highly motivated to access a grooming substrate. *Biology Letters* 14, article 20180303.

McCue M D (2010) Starvation physiology: reviewing the different strategies animals use to survive a common challenge. *Comparative Biochemistry and Physiology A. Molecular and Integrative Physiology* 156, 1–18.

McFarland D (1985) *Problems of Animal Behaviour*. Longmans Scientific and Technical, Harlow.

McFarland D J and Houston A I (1981) *Quantitative Ethology: The State Space Approach*. Pitman Books, London.

McGinnis M Y and Vakulenko M (2003) Characterization of 50-kHz ultrasonic vocalizations in male and female rats. *Physiology and Behavior* 80, 81–8.

McGlone J J (1993) What is animal welfare? *Journal of Agricultural and Environmental Ethics* 6(Suppl 2), 26–36.

McGlone J J (2001) Farm animal welfare in the context of other society issues: towards sustainable systems. *Livestock Production* 72, 75–81.

McGrath N, Dunlop R, Dwyer C, Burman O and Phillips C J C (2017) Hens vary their vocal repertoire and structure when anticipating different types of reward. *Animal Behaviour* 130, 79–96.

McInerney J (1991) A socioeconomic perspective on animal welfare. *Outlook in Agriculture* 20(1), 51–8.

McKean K A, Yourth C P, Lazzaro B P and Clark A G (2008) The evolutionary costs of immunological maintenance and deployment. *Evolutionary Biology* 8, article 76.

Mduma S A R, Sinclair A R E and Hilborn R (1999) Food regulates the Serengeti wildebeest: a 40-year record. *Journal of Animal Ecology* 68(6), 1101–22.

Meehan C L and Mench J A (2007) The challenge of challenge: can problem solving opportunities enhance animal welfare? *Applied Animal Behaviour Science* 102, 246–61.

Meijer J H and Robbers Y (2014) Wheel running in the wild. *Proceedings of the Royal Society* B 281, article 20140210.

Mellen J and McPhee M S (2001) Philosophy of environmental enrichment: past, present and future. *Zoo Biology* 20, 211–21.

Mellor D J (2016a) Moving beyond the 'Five Freedoms' by updating the 'Five Provisions' and introducing aligned 'Animal Welfare Aims'. *Animals* 8(10), article 59.

Mellor D J (2016b) Updating animal welfare thinking: moving beyond the 'Five Freedoms' towards 'a life worth living'. *Animals* 6, article 21.

Mellor D J (2019) Welfare-aligned sentience: enhanced capacities to experience, interact, anticipate, choose and survive. *Animals* 9, 440–55.

Mellor E, Kinkald H M and Mason G (2018) Phylogenetic comparative methods: harnessing the power of species diversity to investigate welfare issues in captive wild animals. *Zoo Biology* 37, 369–8.

Menchetti L, Righi C, Guelphi G, Enas C, Moscati L, Mancini S and Diverio S (2019) Multi-operator qualitative behavioural assessment for dogs entering the shelter. *Applied Animal Behaviour Science* 213, 107–16.

Mendl M (1991) Some problems with the concept of a cut-off point for determining when an animal's welfare is at risk. *Applied Animal Behaviour Science* 31, 139–46.

Mendl M, Burman O H P, Parker R M and Paul E S (2009) Cognitive bias as an indicator of animal emotion and welfare: emerging evidence and underlying mechanisms. *Applied Animal Behaviour Science* 118, 161–81.

Mendl M, Burman O H P and Paul E S (2010) An integrative and functional framework for the study of animal emotion and mood. *Proceedings of the Royal Society* B 277, 2895–904.

Merker B (2007) Consciousness without a cerebral cortex: a challenge for neuroscience and medicine. *Behavioral and Brain Sciences* 30, 63–134.

Meyer-Holtzapfel M (1968) Abnormal behavior in zoo animals. In M W Fox (ed.) *Abnormal Behaviour in Animals*. Saunders, Philadelphia, PA.

Michel M, Beck D, Block N, Blumenfeld H, Brown R, Carmel D et al. (2019) Opportunities and challenges for a maturing science of consciousness. *Nature Human Behaviour* 3, 104–7.

Midgley M (1983) *Animals and Why They Matter. A Journey Around the Species Barrier*. Pelican Books, Harmondsworth.

Millot S, Cerqueira M, Castanheira M F, Overli O, Martins C I M and Oliveira R F (2014) Use of conditioned place preference/avoidance tests to assess affective states in fish. *Applied Animal Behaviour Science* 154, 104–11.

Mitsui, S, Yamamoto M, Nagasawa M, Mogi K, Kikusui T, Ohtani N and Ohta M (2011) Urinary oxytocin as a non-invasive biomarker of positive emotion in dogs. *Hormones and Behavior* 60, 239–43.

Moberg G P (1985) Biological responses to stress: key to assessment of animal well-being? In G P Moberg (ed.) *Animal Stress*. American Philosophical Society, Bethesda, MD, pp. 27–49.

Moberg G P (1987) Problems of defining stress and distress in animals. *Journal of the American Veterinary Medical Association* 19(10), 1201–11.

Moe R O, Stubsjoen S M, Bohlin J, Flø A and Bakken M (2012) Peripheral temperature drop in response to anticipation and consumption of a signaled palatable reward in laying hens (*Gallus domesticus*). *Physiology and Behavior* 106, 527–33.

Moe R O, Nordgreen J, Janczak A M, Bakken M, Spruijt B and Jensen P. (2014) Anticipatory and foraging behaviors in response to palatable food reward in chickens: effects of dopamine D2 receptor blockade and domestication. *Physiology and Behavior* 133, 170–7.

Mogg K, Bradbury K E and Bradley B P (2006) Interpretation of ambiguous information in clinical depression. *Behaviour Research and Therapy* 44, 1411–19.

Monath T P, Kahn L H and Kaplan B (2010) Introduction: One Health perspective. *ILAR Journal* 519(3), 193–8.

Mormède P, Andanson S, Aupérin B, Beerda B, Guémené, Malmkvist J et al. (2007) Exploration of the hypothalamic-pituitary-adrenal function as a tool to evaluate animal welfare. *Physiology and Behavior* 92, 317–39.

Muri K and Stubsjoen S M (2017) Inter-observer reliability of Qualitative Behavioural Assessment (QBA) of housed sheep in Norway using fixed lists of descriptors. *Animal Welfare* 26, 427–35.

Muri K, Stubsjoen S M, Vasdal G, Moe R O and Granquist E G (2019) Association between qualitative behaviour assessments and measures, of leg health, fear and mortality in Norwegian broiler chicken flocks. *Applied Animal Behaviour Science* 211, 47–53.

Murphy S T and Zajonc R B (1993) Affect, cognition and awareness in affective priming with optimal and suboptimal stimulus exposure. *Journal of Personality and Social Psychology* 64, 723–9.

Nakata A, Takahshi M, Irie M and Swanson N G (2010) Job satisfaction is associated with elevated natural killer cell immunity among healthy, white-collar employees. *Brain, Behavior and Immunity* 24, 1268–75.

Nicol C J, Caplen G, Edgar J and Browne W J (2009) Associations between welfare indicators and environmental choice in laying hens. *Animal Behaviour* 78, 413–24.

Nicol C J, Caplan G, Statham P and Browne W J (2011) Decisions about foraging and risk tradeoffs in chickens are associated with individual somatic response profiles. *Animal Behaviour* 82, 255–60.

Niel L and Weary D M (2006) Behavioural responses of rats to gradual-fill carbon dioxide euthanasia and reduced oxygen concentrations. *Applied Animal Behaviour Science* 100, 295–308.

Niel L and Weary D M (2007) Rats avoid exposure to carbon dioxide and argon. *Applied Animal Behaviour Science* 107, 100–9.

Nielsen B L (2018) Making sense of it all; the importance of taking into account the sensory abilities of animals in their housing and management. *Applied Animal Behaviour Science* 205, 175–80.

Nishioka T, Anselmo-Franci J A, Li P, Callahan M F and Morris M (1998) Stress increases oxytocin release within the hypothalamic paraventricular nucleus. *Brain Research* 781, 56–60.

Norwood F B and Lusk J L (2011) *Compassion by the Pound: The Economics of Farm Animal Welfare*. Oxford University Press, Oxford.

Novak J, Stojanovski K, Melotti L, Reichlin T S, Palme R and Würbel H (2016) Effects of stereotypic behaviour and chronic mild stress on judgement bias in laboratory mice. *Applied Animal Behaviour Science* 174, 162–72.

Nussbaum M C (2004) Beyond compassion and humanity. Justice for nonhuman animals. In C S Sunstein and M C Nussbaum (eds) *Animal Rights: Current Debates and New Directions*. Oxford University Press, Oxford, pp. 299–320.

Oatley K and Jenkins J M (1996) *Understanding Emotions*. Blackwell, Oxford.

OIE (World Organisation for Animal Health) (2012) Global Animal Welfare Strategy. OIE, Paris. Available at: http://www.oie.int/fileadmin/Home/eng/Animal_Welfare/docs/pdf/Others/EN_OIE_AW_Strategy.pdf (accessed 8 May 2020).

Olsson I A S, Keeling L J and McAdie T M (2002) The push-door for measuring motivation in hens: laying hens are motivated to perch at night. *Animal Welfare* 11, 11–19.

Olsson I A S and Keeling L J (2005) Why in earth? Dustbathing behaviour in jungle and domestic fowl reviewed from a Tinbergian animal welfare perspective. *Applied Animal Behaviour Science* 93, 259–82.

Olsson T and Sapolsky R (2006) The healthy cortisol response in health and disease. In B B Arnetz and R Ekman (eds) *Stress in Health and Disease*. Wiley-Verlag, Weinheim, pp. 214–25.

Opiol H, Pavlovski I, Michalik M and Mistleberger R E (2015) Ultrasonic vocalizations in rats anticipating circadian feeding schedules. *Behavioural Brain Research* 284, 42–50.

Owen A M (2013) Detecting consciousness: a unique role for neuroimaging. *Annual Review of Psychology* 64, 109–33.

Pacquet P C and Darimont T C (2010) Wildlife conservation and animal welfare: two sides of the same coin? *Animal Welfare* 19, 177–90.

Pajor E A, Rushen J and de Passillé A M B (2003) Dairy cattle's choice of handling treatments in a Y-maze. *Applied Animal Behaviour* Science 80, 93–107.

Palme R (2019) Non-invasive measurement of glucocorticoids: advances and problems. *Physiology and Behavior* 199, 229–43.

Panksepp J (1998) *Affective Neuroscience. The Foundations of Human and Animal Emotions*. Oxford University Press, New York.

Panksepp, J. (2007) Neuroevolutionary sources of laughter and social joy: modelling primal human laughter in laboratory rats. *Behavioral and Brain Research* 182, 231–4.

Panksepp J (2011) The basic emotional circuits of mammalian brains: do animals have affective lives? *Neuroscience and Biobehavioral Reviews* 35, 1791–804.

Passillé A M B de, Christopherson R and Rushen J (1993) Nonnutritive sucking in the calf and postprandial secretion of insulin, CCK and gastrin. *Physiology and Behaviour* 54, 1069–73.

Patterson I J (1965) Timing and spacing of broods in the black-headed gull (*Larus ridibundus* L.). *Ibis* 107, 433–60.

Patterson-Kane E G, Pittman M and Pajor E A (2008) Operant animal welfare: productive approaches and persistent difficulties. *Animal Welfare* 17, 139–48

Paul E S, Harding E J and Mendl M (2005) Measuring emotional processes in animals; the utility of a cognitive approach. *Neuroscience and Biobehavioral Reviews* 29, 469–91.

Pawluski J, Jago P, Henry S, Bruchet A, Palme R, Coste C and Hausberger M (2017) Low plasma cortisol and fecal cortisol metabolite measures as indicators of compromised welfare in domestic horses (*Equus caballus*). *PLoS One* 12, e0182257.

Pennartz C M A, Farisco M and Evers K (2019) Indicators and criteria of consciousness in animals and intelligent machines: an inside-out approach. *Frontiers in Systems Neuroscience* 13, article 25.

Petherick J C, Waddington D and Duncan I J H (1990) Learning to gain access to a foraging and dustbathing substrate by domestic fowl—is out of sight out of mind? *Behavioural Processes* 22, 213–26.

Phillips M L, Williams L M Heining M, Herba C M, Russell T, Andrew C et al. (2004) Differential neural responses to overt and covert presentations of facial expressions of fear and disgust. *NeuroImage* 21(4), 1484–96.

Phythian C J, Michalopoulou E, Cripps P J, Duncan, J S and Wemelsfelder F (2013) On-farm qualitative behaviour assessment in sheep: repeated measurements across time, and association with physical indicators of flock health and welfare. *Applied Animal Behaviour Science* 175(Suppl. 1), 23–31.

Pickering A D, Pottinger T G and Christie P (1982) Recovery of the brown trout, *Salmo trutta* L., from acute handling stress: a time course study. *Journal of Fish Biology* 20, 229–44.

Polgár Z, Blackwell E J and Rooney N J (2019) Assessing the welfare of kennelled dogs—a review of animal based measures. *Applied Animal Behaviour Science* 213, 1–13.

Poirier C and Bateson M (2017) Pacing stereotypies in laboratory rhesus macaques: implications for animal welfare and the validity of neuroscientific findings. *Neuroscience and Biobehavioral Reviews* 83, 508–15.

Porritt F, Shapiro M, Waggoner P, Mitchell E, Thomson T, Nicklin S and Kacelnik A (2015) Performance decline by search dogs in repetitive tasks, and mitigation strategies. *Applied Animal Behaviour Science* 166, 112–22.

Preisser E L, Bolnick D I and Benard M F (2005) Scared to death? The effects of intimidation and consumption in predator-prey interactions. *Ecology* 86, 501–9.

Proctor H (2012) Animal sentience: where are we and where are we going. *Animals* 2, 628–39.

Queenan K, Hasler B and Rushton J (2016) One Health approach to antimicrobial resistance surveillance: is there a business case for it? *International Journal of Antimicrobial Agents* 48, 422–7.

Rabin LA (2003) Maintaining behavioural diversity in captivity for conservation: natural behaviour management. *Animal Welfare* 12, S99–S103.

Ralph C R and Tillbrook A J (2016) The usefulness of measuring glucocorticoids for assessing animal welfare. *Journal of Animal Science* 34(2), 457–70.

Ramp D and Bekoff M (2015) Compassion for a practical evolved ethic for conservation. *BioScience* 65, 323–7.

Rault J-L, Dunshea F and Pluske J (2015) Effects of oxytocin administration on the response of piglets to weaning *Animals* 5, article 037.

Rault J-L, van den Munkhof M and Buisman-Pijlman A (2017) Oxytocin as an indicator of psychological and social well-being in domesticated animals: a critical review. *Frontiers in Psychology* 8, article 1521.

Raz J (1984) *The Morality of Freedom*. Oxford University Press, Oxford.

Regan T (1984) *The Case for Animal Rights.* University of California Press, Berkeley and Los Angeles, CA.

Rehn T, Handlin L, Uvnas-Moberg K and Keeling L J (2014) Dogs' endocrine and behavioural responses at reunion are affected by how the human initiates contact. *Physiology and Behavior* 124, 45–53.

Richards A, French C C, Calder A J, Webb B, Fox R and Young AW (2002) Anxiety-related bias in the classification of emotionally ambiguous facial expressions. *Emotion* 2, 273–28.

Ridley M (1993) *The Red Queen. Sex and the Evolution of Human Nature*. Penguin Books, London.

Roches de A D, Lussert L, Faur M, Herry V, Rainard P, Durand D, Wemelsfelder F and Foucras G (2018) Dairy cows under experimentally-induced *Escheria coli* mastitis show negative emotional states assessed through Qualitative Behaviour Assessment. *Applied Animal Behaviour Science* 206, 1–11.

Rock M L, Karas A Z, Gallo M S, Pritchett-Corning K and Gaskill B N (2014) Housing condition and nesting experience do not affect the Time to Integrate Nest Test (TNT). *Animal Welfare* 23(4), 381–5.

Rollin B (1989) *The Unheeded Cry*. Oxford University Press, Oxford.

Rolls E T (2007) Understanding the mechanisms of food intake and obesity. *Obesity Reviews* 8, 67–82.

Rolls E T (2012) Taste, olfactory, and food texture reward processing in the brain and the control of appetite. *Proceedings of the Nutrition Society* 71, 488–501.

Rolls E T (2014) *Emotion and Decision-Making Explained.* Oxford University Press, Oxford.

Rolls E T (2020) Neural computations underlying phenomenal consciousness: a Higher Order Syntactic Thought theory. *Frontiers in Psychology* 11, article 655.

Roper T J (1976) Self-sustaining activities and reinforcement in the nest building behaviour of mice. *Behaviour* 59, 40–58.

Rosenthal D M (1993) Thinking that one thinks. In M. Davies and G.W. Humphreys (eds) *Consciousness*. Blackwell, Oxford, pp. 197–223.

Rosenthal D M (2005) *Consciousness and Mind.* Oxford University Press, Oxford.

Ross S R, Schapiro S J, Hau J and Lukas K E (2009) Space use as an indicator of enclosure appropriateness: a novel measure of captive animal welfare. *Applied Animal Behaviour Science* 121, 42–50.

Roughan J V, Coulter C A, Flecknell P A, Thomas H D and Sufka K J (2014) The conditioned place preference test for assessing welfare consequences and potential refinements in a mouse bladder cancer model. *PLoS One* 9, article e103362.

Ruegg S R, McMahon B J, Hasler B E, Nielsen R, Rosenbaum L, Speranza C I et al. (2017). A blueprint to evaluate One Health. *Frontiers in Public Health* 5, article 20.

Rushen J (1986a) Aversion of sheep to electro-immobilization and mechanical restraint. *Applied Animal Behaviour Science* 15, 315–24.

Rushen J (1986b) Some problems with the physiological concept of 'stress'. *Australian Veterinary Journal* 63, 359–61.

Rushen J (1991) Problems associated with the interpretation of physiological data in the assessment of animal welfare. *Applied Animal Behavioural Science* 28, 381–6.

Russell J A (2003) Core affect and the psychological construction of emotion. *Psychological Reviews* 110, 145–72.

Russell J A and Barrett L F (1999) Core affect, prototypical emotional episodes, and other things called emotion: dissecting the elephant. *Journal of Personality and Social Psychology* 76, 805–19.

Russo S J, Murrough J W, Han M H, Charney D S and Nestler E J (2012) Neurobiology of resilience. *Nature Neuroscience* 15, 1475–84.

Ryder R (1975) *Victims of Science*. Davis-Poynter, London.

Sand H, Wikenros C, Wabakken P and Liberg O (2006) Effects of hunting group size, snow depth and age on the success of wolves hunting moose. *Animal Behaviour* 72, 781–9.

Sandem A I and Braastad B O (2005) Effects of cow-calf separation on visible eye-white and behaviour in dairy cows. *Applied Animal Behaviour Science* 95, 233–9.

Sandem A I, Braastad B O and Boe K L (2002) Eye white may indicate emotional state on a frustration-contentedness axis in dairy cows. *Applied Animal Behaviour Science* 28, 381–6.

Sapolsky R M (1994) *Why Zebras Don't Get Ulcers: a Guide to Stress, Stress-Related Disease, and Coping*. W. H. Freeman, New York.

Sapolsky R M, Romero L M and Munck A U (2000) How do glucocorticoids influence stress responses? Integrating permissive, suppressive, stimulatory, and preparative actions. *Endocrinological Reviews* 21, 55–89.

Sato W and Aoki S (2006) Right hemisphere dominance in processing of unconscious negative emotions. *Brain and Cognition* 62, 261–6.

Schelonka K, Graulty C, Canseco-Gonzalez E and Pitts M (2017) ERP signatures of conscious and unconscious word and letter perception in an inattentional blindness paradigm. *Consciousness and Cognition* 54, 56–71.

Schrader L and Mueller B (2009) Night-time roosting in the domestic fowl: the height matters. *Applied Animal Behaviour Science* 121, 179–83.

Schraven K, Meder S and Kemper N (2018) The painful face of the horse: the Horse Grimace Scale (HGS) in the practice test. *Tieraerztliche Umschau* 73, 259–64.

Schultz W (2013) Updating dopamine reward signals. *Current Opinion in Neurobiology* 23, 229–38.

Schultz W (2016) Reward functions of the basal ganglia. *Journal of Neural Transmission* 123(7 SI), 679–93.

Schwartz M W and Porte D (2005) Diabetes, obesity, and the brain. *Science* 307, 375–379.

Selye H (1956) *The Stress of Life*. McGraw Hill, New York.

Selye H (1974) *Stress Without Distress*. McClelland and Stewart, Toronto.

Seth A K, Baars B J and Edelman D B. (2005) Criteria for consciousness in humans and other mammals. *Consciousness and Cognition* 14, 119–39.

Sharpe L L, Clutton-Brock T H, Brotherton, P N M, Cameron E Z and Cherry M I (2002) Experimental provisioning increases play in free-ranging meerkats. *Animal Behaviour* 64, 113–21.

Sheriff M J, Dantzer B, Delehamtry R, Palme I and Boonstra R. (2011) Measuring stress in wildlife: techniques for quantifying glucocorticoids. *Oecologia* 166, 869–87.

Sherwin C M, Haug E, Terkelsen M and Vadgama M (2004) Studies on the motivation for burrowing by laboratory mice. *Applied Animal Behaviour Science* 38, 343–58.

Sherwin C M and Nicol C J (1995) Changes in meal patterning by mice measure the cost imposed by natural obstacles. *Applied Animal Behaviour Science* 43, 291–300.

Sherwin C M and Nicol C J (1996) Reorganization of behaviour in laboratory mice, *Mus musculus*, with varying cost of access to resources. *Animal Behaviour* 52, 1087–93.

Shettleworth S J (2010) Clever animals and killjoy explanations in comparative psychology. *Trends in Cognitive Sciences* 14, 477–81.

Singer P (1974) All animals are equal. Reprinted in H Kuhse and P Singer (eds) *Bioethics* (2006). 2nd ed. Blackwell Publishing, Oxford.

Singer P (1976) *Animal Liberation*. Jonathan Cape, London.
Siviy S M and Panksepp J (1985) Energy-balance and play in juvenile rats. *Physiology and Behavior* 35, 435–41.
Sklar A Y, Levy N, Goldstein A, Mandel R, Maril A and Hassin R (2012) Reading and doing arithmetic nonconsciously. *Proceedings of the National Academy of Sciences* 109, 19614–19.
Smid A M C, Weary D M, Costa J H C and von Keyserlingk M A G (2018) Dairy cow preference for different types of outdoor access. *Journal of Dairy Science* 101, 1448–55.
Smith A L and Corrow D J (2005) Modifications to husbandry and housing conditions of laboratory rodents for improved well-being. *ILAR Journal* 46, 140–7.
Smith J H, Wathes C M and Baldwin B A (1996) The preference of pigs for fresh air over ammoniated air. *Applied Animal Behaviour Science* 49, 417–24.
Sneddon L U (2019) Evolution of nocioception and pain: evidence from fish models. *Philosophical Transactions of the Royal Society of London* 374, article 20190290.
Solomon S E, Hughes B O and Gilbert A B (1987) Effect of single injection of adrenalin on shell ultrastructure on a series of eggs from domestic hens. *British Poultry Science* 28, 585–8.
Sotocinal S G, Sorge R E, Zaloum A, Tuttle A H, Martin L J, Wieskopf J S et al. (2011) The Rat Grimace Scale; a partially automated method for quantifying pain in the laboratory rat via facial expressions. *Molecular Pain* 7, article 55.
Špinka, M. (2006) How important is natural behaviour in animal farming systems? *Applied Animal Behaviour Science* 100, 117–28.
Spruijt B M, van den Bos R and Pijlman F T A (2001) A concept of welfare based on reward evaluating mechanisms in the brain: anticipatory behaviour as an indicator for the state of reward systems. *Applied Animal Behaviour Science* 72, 145–71.
Stewart M, Webster J R, Schaefer A L, Cook N J and Scott S L (2005) infrared thermography as a non-invasive tool to study animal welfare. *Animal Welfare* 14, 319–25.
Stoerig, P. (2007) Hunting the Ghost: towards a neuroscience of consciousness. In P D Zelazo, Moscovitch M and Thompson E (eds) *The Cambridge Handbook of Consciousness*. Cambridge University Press, Cambridge, pp. 707–30.
Stranahan A M, Lee K and Mattson M P (2008) Central mechanisms of HPA axis regulation by voluntary exercise. *Neuromolecular Medicine* 10, 118–27.
Stricklin W R, Graves H and Wilson L L (1979) Some theoretical and observed relationships of fixed and portable spacing behaviour of animals. *Applied Animal Behaviour Science* 51, 201–14.
Svennersten K, Nelson L and Uvnäs-Moberg K (1990) Feeding-induced oxytocin release in dairy cows. *Acta Physiologica Scandinavica* 140, 295–6.
Sylvester S P, Mellor D J, Bruce R A and Ward R N (1998) Acute cortisol responses of calves to scoop dehorning using local anaesthetic and/or cautery of the wound. *Australian Veterinary Journal* 76, 118–22.
Takao Y, Okuno Y, Mori Y, Asada M, Yamanishi K and Iso H (2018) Associations of perceived mental stress, sense in purpose in life, and negative life events with the risk of incidental herpes zoster and postherpetic neuralgia. *American Journal of Epidemiology* 187, 251–9.
Tamietto M, Castelli L, Vighetti S, Perozzo P, Geminiani G, Weiskrantz L and de Gelder B (2009) Unseen facial and bodily expressions trigger fast emotional reactions. *Proceedings of the National Academy of Sciences* 106, 17661–6.
Taylor N, Prescott N, Perry G, Potter M, Le Sueur C and Wathes C (1996) Preference of growing pigs for illuminance. *Applied Animal Behaviour Science* 96, 19–31.
Taylor N R, Parker R M A, Mendl M, Edwards S A and Main D C J (2012) Prevalence of risk factors for tail-biting on commercial farms and intervention strategies. *The Veterinary Journal* 194(1), 77–83.

Thayer J F, Aha F, Fredrickson M, Sollers J J iiid and Wager T D (2012) A meta-analysis of heart rate variability and neuroimaging studies: implications for heart rate variability as a marker of stress and health. *Neuroscience and Biobehavioural Reviews* 36, 747–56.

Thornton P D and Waterman-Pearson A E (2002) Behavioural responses to castration in lambs. *Animal Welfare* 11, 203–12.

Thorpe W H (1965) The assessment of pain and distress in animals. Appendix III of *Report of the Technical Committee to Enquire into the Welfare of Animals kept under Intensive Livestock Husbandry Systems*. F W R Brambell (Chairman). Her Majesty's Stationery Office, London. Cmnd. 2836, pp. 71–9.

Toates F (1995) *Stress: Conceptual and Biological Aspects*. John Wiley and Sons, Chichester.

Tononi G (2008) Consciousness as integrated information: a provisional manifesto. *Biological Bulletin* 215(3), 216–42.

Tononi G and Koch C (2015) Consciousness, here, there and everywhere? *Philosophical Transactions of the Royal Society B* 370, 117–34.

Tooley M (2011) Are nonhuman animals persons? In T L Beauchamp and R G Frey (eds) *The Oxford Handbook of Animal Ethics*. Oxford University Press, Oxford, pp. 332–70.

Townsend P (1997) Use of in-cage shelters by laboratory rats. *Animal Welfare* 6, 95–103.

Travain T, Colombo E S, Heinzl E, Bellucci D, Prèvide E P and Valsecchi P (2015) Hot dogs: thermography in the assessment of stress in dogs (*Canis familiaris*). A pilot study. *Journal of Veterinary Behavior—Clinical applications and Research* 10, 17–23.

Travain T, Colombo E S, Silvia E, Grandi L C, Heinzl E, Bellucci D et al. (2016) How good is this food? A study on dogs' emotional responses to a potentially pleasant event using infrared thermography. *Physiology and Behavior* 159, 80–7.

Trivers R L (1974) Parent-offspring conflict. *American Zoologist* 14(1), 249–64.

Troxell-Smith S M, Whelan C J, Magle S B and Brown J S (2017) Zoo foraging ecology: development and assessment of a welfare tool for captive animals. *Animal Welfare* 26, 265–75.

Tye M (2017) *Tense Bees and Shell-shocked Crabs. Are Animals Conscious?* Oxford University Press, Oxford.

Urquiza-Haas E G and Kotrschal K (2015) The mind behind anthropomorphic thinking: attribution of mental states to other species. *Animal Behaviour* 109, 167–76.

Vander A, Sherman J and Lucian, D (1990) *Human Physiology: the Mechanisms of Body Function*. 5th ed. New York, McGraw Hill.

Van Etten K W, Wilson K R and Crabtree R L (2007) Habitat use of red fixes in Yellowstone National Park based on snow tracking and telemetry. *Journal of Mammalogy* 88(6), 1498–507.

Van de Weerd H A, Van Loo P L P, Van Zutphen L F M, Koolhaas J M and Baumans V (1998) Preference for cages with nestboxes. *Animal Welfare* 7, 11–25.

Vaz J, Narayan E J, Kumar R D, Thenmozhi K, Thiyagesan K and Baskaran N (2017) Prevalence and determinants of stereotypic behaviours and physiological stress among tigers and leopards in Indian zoos. *PLoS One* 12, e0174711.

Veasey J S, Waran N K and Young R J (1996) On comparing the behaviour of zoo housed animals with wild conspecifics as a welfare indicator. *Animal Welfare* 5, 13–24.

Veissier I and Miele M (2015) Short historical overview of animal welfare sciences: how a societal concern has become a transdisciplinary subject. *INRA Productions Animales* 28, 399–409.

Walker M and Mason G (2018) Using mildly electrified grids to impose costs on resource access: a potential tool for assessing motivation in laboratory mice. *Applied Animal Behaviour Science* 198, 101–8.

Warburton H and Mason G J (2003) Is out of sight out of mind? The effects of resource cues on motivation in mink *Mustela vison*. *Animal Behaviour* 65, 755–62.

Wathes C M (2010) Lives worth living. *The Veterinary Record* 166, 468–9.

Watters J V (2014) Searching for behavioural indicators of welfare in zoos: uncovering anticipatory behaviour. *Zoo Biology* 33, 251–6.

Weary D M and Fraser D (1995) Calling by domestic piglets—reliable signals of need. *Animal Behaviour* 50, 1047–55.

Weary D M and Robbins J A (2019) Understanding the multiple conceptions of animal welfare. *Animal Welfare* 28, 33–40.

Webb L, Veenhoven R, Harfield J L and Jensen M B (2019) What is animal happiness? *Annals of the New York Academy of Sciences* 438(S1), 62–76.

Webster J (1994) *Animal Welfare. A Cool Eye Towards Eden*. Blackwell Science, Oxford.

Wechsler, B. (1991) Stereotypies in polar bears. *Zoo Biology* 10, 177–88.

Weiskrantz L (1997) *Consciousness Lost and Found*. Oxford University Press, Oxford.

Welfare Quality® (2009) *Welfare Quality® Assessment Protocol for Poultry (Broilers, Laying Hens)*. Welfare Quality® Consortium, Lelystad, the Netherlands.

Welfare Quality® (2018) Homepage. Available at: http://www.EUwelfarequality.net/en-us/home/ (accessed 25 May 2020).

Wells D L (2009) Sensory stimulation as environmental enrichment for captive animals: a review. *Applied Animal Behaviour Science* 118, 1–11.

Welp T, Rushen J, Kramer D L, Festa-Blanchett M and de Passillé A M B (2004) Vigilance as a measure of fear in dairy cattle. *Applied Animal Behaviour Science* 87, 1–13.

Wemelsfelder F, Hunter T E A, Mendl M and Lawrence A B (2001) Assessing the 'whole animal': a free choice profiling approach. *Animal Behaviour* 62, 209–20.

Westerath H, Gygax L and Hillmann E (2014) Are special feed and being brushed judged as positive by calves? *Applied Animal Behaviour Science* 156, 12–21.

Westland K (2014) Training as enrichment and beyond. *Applied Animal Behaviour Science* 152, 1–6.

White T C R (2008) The role of food, weather and climate in limiting the abundance of animals. *Biological Reviews* 83, 227–48.

Widowski T M and Duncan I J H (2000) Working for a dustbath: are hens increasing pleasure rather than reducing suffering? *Applied Animal Behaviour Science* 68, 39–53.

Will A, Wynne-Edwards K, Zhou R and Kitayski A (2019) Of 11 candidate steroids, cortisone concentration standardized for mass in the most reliable steroid biomarker of nutritional stress across feather types. *Ecology and Evolution* 9, 11930–43.

Wood-Gush D G M (1971) *The Behaviour of the Domestic Fowl*. Heinemann Educational Books, London.

Würbel H (2009) Ethology applied to animal ethics. *Applied Animal Behaviour Science* 118, 118–27.

Wyatt, T D (2020) Reproducible research into human chemical communication by cues and pheromones: learning from psychology's renaissance. *Philosophical Transactions of the Royal Society* 375, 20190262.

Yayou K-L, Ito S, Yamamoto N, Kitagawa, S and Okamura H (2010) Relationships of stress responses with plasma oxytocin and prolactin in heifer calves. *Physiology and Behavior* 99, 362–9.

Yeates J W (2017) How good? Ethical criteria for a 'Good Life' for farm animals. *Journal Agricultural and Environmental Ethics* 30, 23–35.

Yeates J W (2018) Naturalness and animal welfare. *Animals* 8, article 33.

Young L J and Zingg H H (2017) Oxytocin. *Hormones, Brain and Behaviour. Volume 3: Molecular and Cellular Mechanisms.* 3rd ed., pp. 259–77.

Zethol T J J, Van Der Heyden J A M, Tolboom J T B M and Olivier B (1994) Stress-induced hyperthermia in mice. A methodological study. *Physiology and Behavior* 55, 109–15.

Zupan M, Buskas J, Altimiras J and Keeling L J (2016) Assessing positive emotional states in dogs using heart rate and heart rate variability. *Physiology and Behavior* 155, 102–11.

Index